Pesticide Safety

A Study Manual for Private Applicators

Third Edition

University of California
Statewide Integrated Pest Management Program
Agriculture and Natural Resources
Davis, California

Publication 3383-3

To order or obtain UC ANR publications and other products, visit the UC ANR online catalog at https://anrcatalog.ucanr.edu/ or phone 1-800-994-8849. Direct inquiries to

University of California
Agriculture and Natural Resources
Communication Services
2801 Second Street
Davis, CA 95618

Telephone 1-800-994-8849
E-mail: anrcatalog@ucanr.edu

Third Edition, 2021

Publication 3383-3

ISBN-13: 978-1-62711-132-4

Library of Congress Cataloging-in-Publication Data

Names: Whithaus, Shannah M., author. | Blecker, Lisa A., editor. | University of California Integrated Pest Management Program.
Title: Pesticide safety : a study manual for private applicators / Shannah M. Whithaus, Lisa A. Blecker.
Other titles: Publication (University of California (System). Division of Agriculture and Natural Resources) ; 3383.
Description: Third edition. | Oakland, California : University of California Statewide Integrated Pest Management Program, Agriculture and Natural Resources, 2021. | Series: Publication / University of California, Statewide Integrated Pest Management Project, Division of Agriculture and Natural Resources ; 3383 | Includes bibliographical references and index. | Summary: "This manual is for all California farm owners, managers, and employees involved with the use of pesticides. It covers pesticide labels, mixing and applying pesticides, human and environmental hazards of pesticide use, and pesticide emergencies"-- Provided by publisher.
Identifiers: LCCN 2020056562 | ISBN 9781627111324 (paperback)
Subjects: LCSH: Pesticides--Safety measures--Handbooks, manuals, etc. | Pesticides--Application--Safety measures--Handbooks, manuals, etc.
Classification: LCC SB952.5 .W45 2021 | DDC 363.738/498--dc23
LC record available at https://lccn.loc.gov/2020056562

Editing and project management: Linda Ribera. Design: Will Suckow. Based on original design: Celeste Rusconi. Archivist: Evett Kilmartin. Illustrations by UC ANR staff, except as noted in the captions. Cover photos: Marty Martino (top); Evett Kilmartin (bottom). All other photos by Jack Kelly Clark, except as follows: GoatThroat Pumps: 7-21A; Anna K. Hunter: 4-8, 6-1, 6-5, 9-10, 12-4; Kenzo Estate Winery (Napa, CA): 11-6; Petr Kosina: 7-9, 7-10; Marty Martino: 8-1; Suzanne Paisley: 5-9; Cheryl A. Reynolds: 7-6, 7-12, 10-4, 12-3; David Rosen: 1-2; Shannah M. Whithaus: 7-2, 7-3; The Tree Center: 2-10

UC PEER REVIEWED This publication has been peer reviewed for technical accuracy by University of California scientists and other qualified professionals. This review process was managed by UC ANR Associate Editor for Urban Pest Management David Haviland.

Printed in United States on recycled paper.

6m-pr-7/21-LR/WS

Contents

Contributors and Acknowledgments v
Introduction vi
1. Pest Management 1
Integrated Pest Management 2
Why Monitoring for Pests Is Important 3
Creating an Effective IPM Program 4
Chapter 1 Review Questions 7
2. Pest Identification 9
Understanding Pest Biology 10
Identifying Pests 10
Common Pests in California 12
Chapter 2 Review Questions 28
3. Pesticides 29
Pesticides and Their Hazards 30
How Pesticides Are Organized 32
Adjuvants 38
Chapter 3 Review Questions 39
4. Laws and Regulations 41
Pesticide Registration and Labeling 42
Chapter 4 Review Questions 53
5. Environmental Hazards 55
The Environment and the Hazards of Pesticide Use 56
Pesticide Characteristics 56
How Pesticides Behave in the Environment 57
Environmental Impacts of Pesticide Application 61
Chapter 5 Review Questions 64
6. Human Hazards 67
Potential for Human Injury 68
Harmful Effects of Pesticide Exposure 72
Other Problems Associated with Pesticides 74
Chapter 6 Review Questions 75
7. Personal Protective Equipment and Personal Safety 77
Keeping People Safe 78
Personal Protective Equipment 80
Engineering Controls 90
Chapter 7 Review Questions 93
8. Using Pesticides Safely 95
Pesticide Applicator Safety 96
Safe Application Methods 104
Cleaning Application Equipment 106
Personal Cleanup 106
Chapter 8 Review Questions 107

9. Application Equipment 109
Application Equipment 110
Application Methods and Equipment 117
Maintaining Application Equipment 117
Chapter 9 Review Questions 125
10. Calibrating Pesticide Application Equipment 127
Why Calibration Is Essential 128
Equipment Calibration Methods 128
Calculation for Active Ingredient, Percentage Solutions, and Parts per Million Dilutions 146
Using System Monitors and Controllers 149
Chapter 10 Review Questions 151
11. Using Pesticides Effectively 153
Predicting Pest Problems 154
Making Pesticide Use Decisions 154
Choosing the Right Pesticide 155
Applying Pesticides Effectively 158
Mixing Pesticides 159
Pesticide Resistance 161
Preventing Offsite Movement of Pesticides 162
Chapter 11 Review Questions 170
12. Pesticide Emergencies and Emergency Response 173
First Aid 174
If Pesticides Get on Your Skin or Clothing 176
If Pesticides Get in Your Eyes 177
If Pesticides Are Inhaled 177
If Pesticides Are Swallowed 178
Pesticide Leaks and Spills 179
How to Deal with a Pesticide Fire 182
How to Deal with Stolen Pesticides 182
Misapplication of Pesticides 183
Reviewing Emergency Response to Accidents 184
Chapter 12 Review Questions 185
Appendix A 187
Appendix B 188
Appendix C 189
Appendix D 196
Answers to Review Questions 199
Glossary 201
References 221
Index 225

Contributors and Acknowledgments

This manual was prepared by the University of California Agriculture and Natural Resources Statewide Integrated Pest Management (IPM) Program under a memorandum of understanding with the California Department of Pesticide Regulation (DPR).

Prepared by the University of California

Statewide IPM Program at Davis
Shannah M. Whithaus, Writer/Editor
Lisa A. Blecker, Technical Editor
Tunyalee Martin, Associate Director of Communications
James J. Farrar, Director, Statewide IPM Program at Davis
David Haviland, Associate Editor for Pest Management, ANR Communication Services

Technical Advisory Committee and Principal Contributors

The following people provided ideas, information, and suggestions and reviewed manuscript drafts:
Scott Bowden, Deputy Agricultural Commissioner, Sutter County
Matt Bozzo, Farm Manager, Golden Gate Hop Ranch, Yuba City
Laurie Brajkovich, Department of Pesticide Regulation
Jose Chang, Deputy Agricultural Commissioner (Pesticides), Napa County
Ruth M. Dahlquist-Willard, Small Farms and Specialty Crops Farm Advisor, Fresno and Tulare Counties
Joseph Damiano, Department of Pesticide Regulation
Surendra Dara, University of California Cooperative Extension Advisor, San Luis Obispo
Franz Niederholzer, University of California Cooperative Extension Advisor, Sutter-Yuba Counties
David Ogilvie, Director of Production, Wilson Farms, Clarksburg
Leslie Talpasanu, Department of Pesticide Regulation
Scott Thomsen, Department of Pesticide Regulation
Alya Wakeman-Hill, Agricultural/Standards Specialist, Fresno County Department of Agriculture
Lynn Wunderlich, University of California Cooperative Extension Advisor, Placerville
John Young, Agricultural Commissioner/Sealer, Yolo Certified Organic Agriculture, Yolo County Department of Agriculture
Nick Zanotti, Farm Manager, Bypass Farms, West Sacramento
Amanda Zito, Supervising Ag/Standards Specialist, Sanger District Office, Fresno County Department of Agriculture

Additional contributions by

Anna Katrina Hunter, Pesticide Safety Education Program Writer
Alana Mari Schoen, Pesticide Safety Education Program Intern

Introduction

As an employee, owner, leaseholder, or manager who handles pesticides on a California farm, you are required to *know*, *understand*, and *follow* federal, state, and local pesticide regulations if any pesticides are used on that farm. To help you protect yourself and other employees, you should regularly review *Pesticide Safety Information Series A-1* through *A-11* leaflets, which are available at your local county agricultural commissioner's office or on the Department of Pesticide Regulation (DPR) website, cdpr.ca.gov/docs/whs/psisenglish.htm. These materials are updated regularly to reflect the latest changes to pesticide-related laws and regulations. Most importantly, federal and state laws require that anyone using California restricted materials be a *certified applicator*. Even when a certified commercial applicator is hired to apply restricted materials on the farm where you work, someone working for the farm operation (an employee, manager, leaseholder, or the owner) must be certified as a private applicator.

Handling pesticides requires many special skills and responsibilities. If you handle pesticides, you need to recognize their hazards and how to avoid them. You must also be familiar with all local, state, and federal laws regulating the sale, use, storage, transportation, application, and disposal of pesticides used on the farm where you work. If you supervise pesticide handlers, you are responsible for seeing that these employees handle and use pesticides properly and safely. A pesticide handler is any person who

- handles opened pesticide containers
- mixes, loads, or applies pesticides or assists in application activities (such as flagging)
- incorporates pesticides into soil
- adjusts, repairs, or removes treatment site coverings
- enters treated areas during an application or before the inhalation exposure level listed on pesticide product labeling has been reached or greenhouse ventilation criteria have been met
- cleans, maintains, services, repairs, or otherwise handles equipment that could contain pesticide residues

To use restricted materials, you must demonstrate, through an examination process, that you can competently and safely handle these especially hazardous chemicals. Once you successfully pass California's **Private Applicator Certification (PAC) examination** and meet other requirements, you may apply for a permit to purchase, possess, and use California restricted materials for application to agricultural areas on the farm where you work. Staff at the agricultural commissioner's office where you receive the permit will tell you how long the permit will remain valid. You will need to complete 6 hours of continuing education or take the certification examination every 3 years to maintain your PAC before a new permit is issued. Your PAC only allows you to make agricultural applications. To make postharvest or other nonagricultural pesticide applications, you will need to become certified in the DPR category that corresponds with that type of application.

This updated study guide reflects changes in laws, regulations, the scientific understanding of our environment and its ecosystems, and application technology. It will provide you with the core knowledge you must have in order to become a certified private applicator. Its purpose is to help you learn safe and effective ways of using pesticides on the farm where you work. It describes how to prevent accidents and how to avoid injury and environmental problems. If you would like more information on any of the topics treated in *Pesticide Safety for Private Applicators, 3rd Edition*, you can read *The Safe and Effective Use of Pesticides, 3rd Edition*, which is Volume 1 in the Pesticide Application Compendium series (not required for the PAC examination).

How to Use This Book

Read this book carefully to prepare for the PAC examination. DPR uses this test to certify farm owners, leaseholders, and managers who may have to purchase restricted materials, as well as farm employees who supervise pesticide handlers or will be training handlers and fieldworkers to work safely around pesticides. A list of knowledge expectations (descriptions of what you should know after reading the chapter) are given at the beginning of each chapter to guide you as you study. Individual knowledge expectations appear alongside relevant content throughout each chapter, which will help you focus on the information that is most likely to appear on the examination. Please note that knowledge expectations are placed closest to the most detailed information on that subject, and that additional information on that subject will not be marked. Other sections of the chapter will likely cover information relevant to any of the knowledge expectations, so you should read all sections of the text, not just the ones where a knowledge expectation appears.

Additional information you may wish to review will be available in the appendices. Information contained there will not appear on any certification exam, but may be useful to you now or in the future.

What DPR Is Testing

DPR exams assess your competence in the handling and/or supervising the handling of restricted-use pesticides. Questions on the exams are similar to the review questions at the end of each of this book's chapters. You will be tested on information related to the knowledge expectations provided at the beginning and throughout each chapter to ensure that you know how to safely and effectively handle or supervise the handling of restricted-use pesticides according to California law.

Using the Review Questions

The review questions at the end of each chapter are there to test your grasp of the knowledge expectations in that chapter. Begin your study of each chapter by reading through the knowledge expectations and the review questions. Take note of the material you do not fully understand. Then, review the chapter to locate the sections that deal with that information. Read those sections carefully and then review the rest of the chapter so that you fully understand each concept covered by the knowledge expectations—even the ones you feel you already understand.

When you finish studying, answer the review questions. Check your answers with the correct answers in the "Answers to Review Questions" at the end of the book. If you missed any of the questions, go back and reread the parts of the chapter that cover that information.

Useful Resources

Besides this text, there are two important sources you should rely on for information regarding pesticides and pest management.

- County agricultural commissioners (CACs) are DPR regulatory officials. Their offices throughout the state have the responsibility, among other functions, for issuing permits for restricted-use pesticides; monitoring pesticide use, storage, and disposal; and enforcing pesticide laws and regulations. Agricultural commissioners' offices provide local information on pesticide use, storage, transportation, disposal, and hazards. Contact your local CAC's office if there is any pesticide emergency.
- The University of California, through its Cooperative Extension program, maintains offices in most of California's counties. Experts who staff these offices are able to help you locate pest identification, pest management, and pesticide use information you need, or they can access a network of University of California scientists when additional assistance is needed. You can also access a wealth of pest management information through the UC IPM website: ipm.ucanr.edu.

Chapter 1
Pest Management

Integrated Pest Management 2
Why Monitoring for Pests Is Important 3
Creating an Effective IPM Program 4
Chapter 1 Review Questions 7

Knowledge Expectations

1. Explain how using integrated pest management can help you make pesticide applications more effective.
2. Explain how secondary pest outbreaks can occur.
3. Describe the ways that monitoring can help you make a more informed pesticide application.

Integrated Pest Management

Explain how using integrated pest management can help you make pesticide applications more effective.

Integrated pest management (IPM) is a pest management program that uses knowledge of pest biology and extensive monitoring to understand a pest and its potential for causing economic injury. IPM combines a variety of control methods to prevent pest damage sustainably, including prevention, cultural practices, exclusion, use of natural enemies, host plant resistance, and pesticide applications. The goal is to achieve long-term pest suppression with minimal impact on people, nontarget organisms, and the environment.

The key to using IPM methods successfully is correct identification of the pest to be managed. Monitoring, sampling, knowing common pests at the site, and using identification keys can help with this task. Another early step that can improve your pest control efforts is finding out whether the identified organism is a key pest or a secondary pest.

Explain how secondary pest outbreaks can occur.

- Key pests are those that cause major damage on a regular basis unless you successfully manage them. Many weeds are key pests because they compete with crop plants for resources. These weeds require regular control efforts to prevent damage.
- Secondary pests are those that become problems only after a key pest is controlled. For example, some weed species become pests after the more competitive weeds have been controlled. And certain plant-feeding pests begin to cause damage once the key pest is under control or after pesticides applied to the site kill their natural enemies.

You are probably familiar with the key pests affecting crops you grow, but you may be less familiar with possible secondary pests. For help dealing with secondary pest outbreaks, check the UC IPM Pest Management Guidelines and year-round IPM programs for information.

CASE STUDY 1-1

Biological Control in Vicente's Vineyards

Vicente, a farmer in the San Joaquin Valley, noticed an increase in the number of Argentine ants around the vines in the southwestern quadrant of vineyard 12-A. Following that discovery, he started seeing waxlike honeydew deposits on cordons and grape clusters (Fig. 1-1). After searching the leaves, cordons, and clusters, he took several samples of an insect he soon identified as vine mealybug. He then scouted his other vineyards to make sure there were no other areas of infestation on the property. After reading the UC IPM Pest Management Guidelines for vine mealybug on grapes, he decided to treat the initial infestation with a pesticide and to follow this treatment with two other control methods: biological control and cultural control.

FIGURE 1-1.

(A) Argentine ant tending mealybugs. (B) A fieldworker uses a hand lens to examine a grape leaf for beneficials. (C) Grape mealybug mobile first-instar nymphs or crawlers and honeydew.

Vicente began his biological control program with augmentation, buying and releasing mealybug destroyers into the affected vineyard early in the season. Releasing additional predators when mealybug eggs and crawlers were present helped suppress the mealybug population, and it reduced the number of pesticide applications needed for control at that time.

In order to increase his chances of successful control later in the season, Vicente decided to use classical biological control. He bought and released an imported wasp, *Anagyrus pseudococci*, that parasitizes this species of mealybug late in the growing season. By combining these two forms of biological control, Vicente was able to extend the effectiveness of his initial pesticide application without further chemical control.

In addition, Vicente learned that mealybug nymphs and females cannot fly, so he asked everyone to use the wash station he had set up near the affected area after they finished working there to keep them from carrying insects away from the infested area. The station contained equipment such as a power washer for rinsing dust or mud off equipment and blowers to help remove dust from clothing, tools, and personal protective equipment (PPE). Hopefully, he explained to the fieldworkers, taking the extra time to clean equipment and clothing will reduce the need to spray the other vineyards on the property with pesticides.

With the cooperation of his employees, Vicente combined various biological controls with reasonable sanitation methods and insecticides to contain and suppress the mealybugs well enough to prevent damage to his grapes.

Why Monitoring for Pests Is Important

Describe the ways that monitoring can help you make a more informed pesticide application.

Monitoring is the use of specific procedures to watch the activities, growth, and development of pests over a period of time. It is your key to building an effective pest management program. Detection tells you whether pests are present or not and helps you anticipate pest outbreaks. Continued monitoring provides important information about a pest's life stages and habits as well as changes at the site that can affect the outcome of pest control measures (Fig. 1-2). Establishing a monitoring program not only allows you to detect pests, it also lets you

- observe seasonal changes in pest populations
- track natural enemy populations
- choose the most effective pesticide
- time pesticide applications properly
- assess the effectiveness of control measures

Monitoring is fundamental to IPM and is essential for effective decision making.

FIGURE 1-2.

Kern County Farm Advisor David Haviland is visually monitoring pests on navel orange leaves.

Follow-Up Monitoring

Monitor after every treatment to learn whether the control activity was successful. Follow-up monitoring includes checking a site even if no treatment has been recommended; if you decide not to treat, you should continue follow-up monitoring until the pest is no longer a threat to the crop. After treatment, monitor for signs that you have gotten the level of control you wanted from the application. For instance, after a fungicide application, perform follow-up monitoring to verify that coverage was thorough and the pathogen was suppressed. Also, if the conditions for disease development continue, keep monitoring to determine whether further management actions are required as unprotected foliage becomes susceptible.

If a pesticide application was unsuccessful, regular follow-up monitoring and recordkeeping allow you to assess the problem(s) and institute corrective measures or change how you make the next application.

Creating an Effective IPM Program

IPM programs use methods that keep pests from becoming a problem, such as growing healthy plants that can withstand pest attacks, using disease-resistant plants, and encouraging populations of predatory or parasitic insects that help keep pest populations from increasing. An effective IPM program combines biological control, chemical control, mechanical and physical controls, and cultural controls, which are covered in the next sections. Sidebar 1-1 contains a sample plan that you can use to start your own IPM program.

Biological control is the use of living biological control agents or natural enemies—predators, parasites, pathogens, and competitors—to control pests and their damage (Fig. 1-3). Invertebrates, plant pathogens, nematodes, weeds, and vertebrates have many natural enemies. Types of biological control include

- classical biological control (importing natural enemies to control invasive species)
- augmentation (raising and releasing native natural enemies to control local pests)
- naturally occurring control (protecting native natural enemies to control local pests)

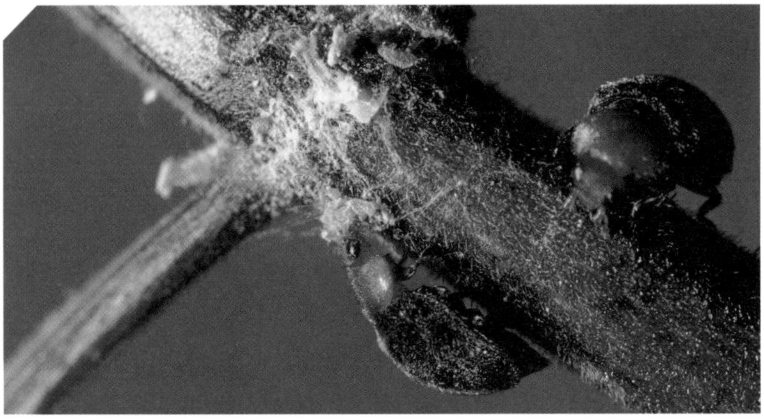

FIGURE 1-3.

Adult mealybug destroyer beetles attacking soft-bodied mealybugs.

SIDEBAR 1-1

Sample IPM Plan

1. Background and site information
    - farm location and mailing address
    - site overview and history (past plantings and pest management operations)
    - resource concerns (positives and negatives of the site)
    - orchard/field maps and descriptions
2. Environmental risk assessment and mitigation
    - soils description and map
    - sensitive area and buffer zone map
    - current management practices
    - pesticide resistance concerns and management
    - mitigation practices to reduce environmental risk in
        - field 1
        - field 2, etc.
        - remaining nonagricultural lands and farmstead
    - pesticide storage, mixing, and container disposal
    - emergency action plan and restricted-entry interval tracking
    - implementation records
    - additional comments
3. Scouting and monitoring guidelines
    - pest history
    - list of crops to be maintained
    - goals for pest management
    - scouting methods for arthropods (both beneficial and pest organisms), disease, weeds, vertebrates, and location of recorded results
    - weather forecasting
    - methods for continued monitoring of pests and location of recorded results
4. Relevant pest management methods
    - list combined pest management methods for use on
        - nonagricultural lands and farmstead
        - field 1
        - field 2, etc.

CASE STUDY 1-2

Conserving Natural Enemies in Marisol's Mandarins

Marisol discovered scabby, silvery scars in rings on some of the mandarins in her orchard. To identify the pest responsible, she checked the UC IPM Pest Management Guidelines (PMGs) and spoke to her local Cooperative Extension advisor, who confirmed Marisol's identification of citrus thrips (Fig. 1-4) and told her to monitor the orchard for them. Marisol recorded high numbers of thrips but also noticed lots of predatory mites and lacewings, both natural enemies of citrus thrips.

Armed with the data she collected from her monitoring efforts, Marisol chose to treat the thrips with spinosad, which the Citrus Thrips PMG said would have fewer harmful effects on the natural enemy populations she observed. She made the application using an "outside coverage" method that targeted the areas most likely to experience scarring.

A few days after making the application, Marisol came back to the orchard to see if it had been effective. She noticed a reduction in the number of thrips on the fruit and saw that a good number of predatory mites and lacewings survived to control the remaining thrips. After several weeks, damage from citrus thrips no longer threatened to ruin her mandarins, and Marisol did not have to use any additional insecticides that season.

FIGURE 1-4.

Citrus thrips like this one are often found on flowers, leaves, and fruits.

Chemical control is the use of naturally occurring or synthetic pesticides. In IPM, pesticides are used only when needed and in combination with other approaches for more effective, long-term control. Also, pesticides are selected and applied in a way that minimizes their possible harm to people and the environment. With IPM, you'll use the most selective pesticide that will do the job and be the safest for other organisms and for air, soil, and water quality. For example, spot-spraying a few weeds instead of an entire area would be the IPM approach.

Mechanical and physical controls kill a pest directly or make the environment unsuitable for it. Traps for rodents are examples of mechanical controls (Fig. 1-5). Physical controls include mowing for weed management, steam sterilization of the soil for disease management, and using barriers such as netting or cloth mesh to keep birds or insects out (exclusion).

Cultural controls are practices that reduce pest establishment, reproduction, dispersal, and survival. For example, changing irrigation practices can reduce pest problems, since

FIGURE 1-5.

A conibear trap is set and placed over the entrance to a ground squirrel burrow, and is secured with a stake.

too much water can increase root disease and weeds, and crop rotation can be used to reduce nematode populations. Cultural controls include

- site selection (choosing a site that is best for the crop plant)
- sanitation (removing weeds before they go to seed, destroying diseased plant materials, cleaning plows and other farming tools, etc.)
- habitat modification (removing weeds from around fields to eliminate shelter for rodents, leaving a 6- to 8-inch pad of dry manure in poultry houses for natural enemies of flies, etc.)
- host resistance (breeding or selecting plants and animals that can resist pest attacks)
- planting date (planting earlier or later in the season to avoid times when pests are a problem)

Chapter 1 Review Questions

1. The term used to describe a pest that becomes a problem after the main pest is controlled is _______.

☐ a. an occasional pest
☐ b. a secondary pest
☐ c. a minor pest

2. How did Marisol's monitoring program affect her choice of control methods?

☐ a. She applied pesticide to keep pests below damaging levels immediately, then made another application later in the season to maintain control.
☐ b. She decided against using any pesticide because she saw that natural enemies were keeping the pest under sufficient control.
☐ c. She chose a pesticide and an application method that would help preserve natural enemies for longer-term control.

3. Which of the following activities are part of an IPM program? Select all that apply.

☐ a. eliminating all insects present in an area
☐ b. identifying pests accurately
☐ c. preventing pest problems
☐ d. removing vegetation completely in an area
☐ e. monitoring for pests and pest damage
☐ f. combining pest management tools
☐ g. applying the same pesticide several times each season

4. How did adding sanitation methods (such as cleaning stations) to his IPM program help Vicente keep his mealybug problem from spreading?

☐ a. It kept his workers from accidentally bringing mealybugs into unaffected vineyards on their equipment or clothing.
☐ b. It stopped workers from completing tasks too quickly and possibly leaving areas before pests were fully eliminated.
☐ c. It helped workers eliminate the need to apply pesticides in other vineyards by preserving natural enemies.

5. Why is it important to monitor the life stage of pests you are trying to control?

☐ a. A restricted materials permit won't be issued if you can't identify a pest by its life stage.
☐ b. Certain pests at certain life stages are best to avoid, even when they are damaging crops.
☐ c. The pesticide product must be applied at the right time in order to adversely affect pests.

6. Monitor after every treatment to learn _________.

☐ a. whether the control activity was successful
☐ b. how much residue remains on leaves
☐ c. if people working at the site are wearing the correct personal protective equipment

CHAPTER 2
PEST IDENTIFICATION

Understanding Pest Biology 10
Identifying Pests 10
Common Pests in California 12
Chapter 2 Review Questions 28

Knowledge Expectations

1. Explain why understanding pest biology is important when managing pests.
2. Explain why identifying pests correctly is important.
3. List and describe the types of resources and references available for identifying pests, symptoms of infestation, and damage caused by pests.
4. List examples of common pests in California from each main group (weeds, arthropods, pathogens, vertebrates), and describe the damage they cause.

Understanding Pest Biology

Explain why understanding pest biology is important when managing pests.

Explain why identifying pests correctly is important.

Before trying to control a pest, you must identify it and understand its biology. Be certain any injury or damage you see is actually caused by the pest you have identified and cannot be blamed on anything else. Once you have identified the pest and its damage, learn about its life cycle and growth, as well as the life cycle and growth of the crop or animal host. Then use this information to form your pest control plans. Misidentification, lack of information about a pest, and poor understanding of a plant's, animal's, or crop's most vulnerable growth stages could cause you to choose the wrong control method or apply the control at the wrong time. Making poor choices based on poor information is the most common cause of pest control failure.

This chapter reviews some of the ways to identify pests, and discusses the four main groups of common pests:

- weeds: unwanted plants
- invertebrates: insects, mites, and their relatives; nematodes and other microscopic worms; and snails and slugs
- vertebrates: birds, reptiles, amphibians, fish, and mammals
- pathogens: bacteria, viruses and viroids, fungi, phytoplasmas, and other microorganisms

Identifying Pests

List and describe the types of resources and references available for identifying pests, symptoms of infestation, and damage caused by pests.

You can identify pests by using the pictures and descriptions included in this chapter and looking at identification guides or the University of California's print and online resources. Or, a specialist can examine and identify the pest for you. When having pests identified, always collect several specimens to represent different life stages or various symptoms of damage.

What makes most mites, nematodes, and plant pathogens hard to identify in the field is their small size. Accurate identification requires the use of a hand lens or microscope, special tests, or careful analysis of damage. Other diagnostic tools are needed to decide whether the damage you see is caused by pests or abiotic (nonliving) factors. Often the pest's host and location are important to making positive identifications. Information on the environmental conditions where you collect pests and the time of year that you collect them give clues to the pest's identity. Resources on the UC IPM website, ipm.ucanr.edu, can help you link hosts with their most common pests as well as the time of year those pests will be most numerous.

Pest species may look different depending on life cycle or time of year. Small weed seedlings, for example, do not always look like mature weeds. The way many insect species look changes as they develop from eggs through adult stages. And some endangered or beneficial species may look like pests depending on their life stage, so it is critical to identify pests correctly.

Identification Experts

Only trained experts using special methods and equipment can positively identify pests such as nematodes and most pathogens. Identification resources for these and other pests are available (typically for a fee) from private laboratories, pest control companies, veterinarians, and licensed pest control advisers. You can also use identification resources provided at no or low cost from the University of California or state government agencies. These include

- local UC Cooperative Extension (UCCE) advisors
- local County Agricultural Commissioner's offices
- California Department of Food and Agriculture's Plant Pest Diagnostics Center

Identification Keys

Identification keys describe individual pests to help you figure out which one is causing problems. Unless you know the terms used to describe a pest's physical structures, many keys are hard to use because they are created by and for experts. However, simple keys are usually available for common pests. A dichotomous key consists of a series of paired statements. To use such a key,

SIDEBAR 2-1

Links to Pest Identification Publications Available from the University of California

FREE PUBLICATIONS

- Pest Management Guidelines for Agricultural Pests (identification keys are linked in some of these pages): ipm.ucanr.edu/agriculture/
- Weed Gallery: ipm.ucanr.edu/PMG/weeds_intro.html
- Natural Enemies Gallery: ipm.ucanr.edu/PMG/NE/index.html

PAID PUBLICATIONS

Pest Identification Card Sets:

- *Pests of Vegetable Crops* (ANR Publication 3553): ipm.ucanr.edu/IPMPROJECT/vegetablesidcards.html
- *Pests of Vineyards* (ANR Publication 3532, also available in Spanish): ipm.ucanr.edu/IPMPROJECT/ADS/vineyardidcards.html
- *Pests of Tree Fruit* (ANR Publication 3426): anrcatalog.ucanr.edu/Details.aspx?itemNo=3426

Pest Management Books and Publications:

- ipm.ucanr.edu/IPMPROJECT/pubs.html

choose a statement from the first pair that best fits the pest you saw or collected at the site. The statement you choose will lead you to another pair of statements. Continue working through the paired statements in this manner until the key leads you to the pest's identity. Dichotomous keys mainly use structural features, but sometimes they rely on the organism's color or size, especially with weeds. Keys may include photographs or drawings to help illustrate features described in the key. Table 2-1 is an example of just one of the dichotomous keys that can be found on the UC IPM website (see Sidebar 2-1 for additional UC publications used for pest identification).

TABLE 2-1:

Dichotomous key for grasses

Characteristics	Species
1a. New leaves are rolled in the bud. 1b. New leaves are folded in the bud.	Go to line 2. Go to line 5.
2a. Auricles are present. 2b. Auricles are absent.	Go to line 4. Go to line 3.
3a. Ligules are present. 3b. Ligules are absent.	Go to line 5. barnyardgrass
4a. Auricles are variable in size, clasping to blunt; leaf blades appear glossy, shiny. 4b. Auricles are long and clasping; leaf blades are hairy, giving a soft, velvety look.	Italian ryegrass hare barley
5a. Ligule is membranous. 5b. Ligule is a fringe of hairs.	Go to line 6. yellow foxtail
6a. Blade and sheath are without hairs. 6b. Blade and/or sheath are very hairy or have fine, short hairs.	bentgrasses Go to line 7.
7a. Leaf blade is hairy on both surfaces; stiff, perpendicular hairs on sheath. 7b. Leaves and sheath are sometimes hairy but mostly smooth or with fine hairs.	large crabgrass Go to line 8.
8a. Ligule is toothed, <1/25> to <1/5> inch (1–5 mm) long. 8b. Ligule is rounded and somewhat wavy, <1/25> to <1/12> inch (1–2 mm) long. 8c. Ligule is rounded or with a blunt tip, <1/12> to <1/3> inch (2–8 mm) long.	German velvetgrass smooth crabgrass dallisgrass

Photographs and Drawings

Whenever possible, use photographs and drawings to help with identification because they provide good visual information about the pest (Fig. 2-1) and its damage, and they can help you locate the pest's unique features. Use publications such as the California Department of Food and Agriculture's *Vertebrate Pest Control Handbook* (available online from the Vertebrate Pest Control Research Advisory Committee's website) and the University of California publications *Weeds of California and Other Western States* and UC IPM manuals. Sidebar 2-1

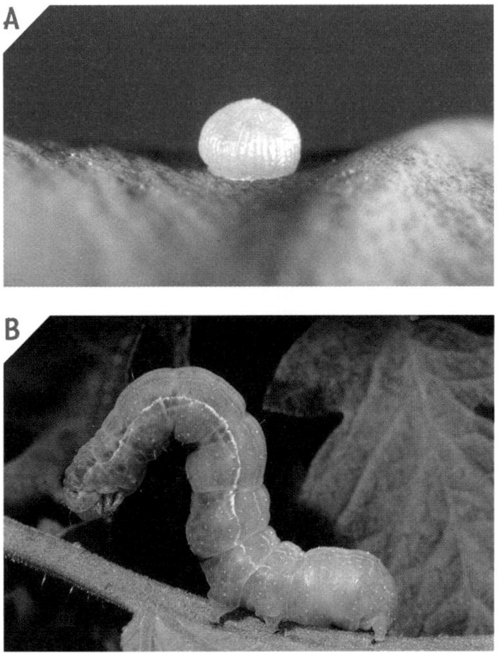

FIGURE 2-1.

Photographs such as these of a cabbage looper egg (A) and larva (B) show unique physical characteristics or coloration patterns that are useful identification aids.

contains links to additional UC resources, such as weed gallery pages, Pest Management Guidelines, and IPM manuals for various crops that can help you identify and manage pests.

Characteristic Signs

Pests may leave signs that help you identify them. Birds and rodents build nests that are often unique to a species. Trails in grass or tracks in dirt are helpful clues to rodent identification. Rodent fecal pellets and insect frass are also important identification aids. Rodents and other mammals dig distinctive burrows in the ground and often leave identifying gnaw marks on tree trunks or other objects, or from feeding. The marks left by feeding can help you identify many insects, as well. Weeds may have unique flowers, seeds, fruits, or unusual growth habits. You can also look for remains of weed plants from last season. Fungi and other pathogens sometimes cause specific types of damage, deformation, or color changes to leaves, fruits, or other plant parts.

Common Pests in California

List examples of common pests in California from each main group (weeds, arthropods, pathogens, vertebrates) and describe the damage they cause.

Weeds

Weeds compete with crops for water, nutrients, light, and space, and they may also interfere with farming operations. Some weed species can poison livestock. Others release compounds into the soil that slow or stop the growth of other plants. Weeds clog irrigation canals and drainage ditches. Uncontrolled weeds contaminate products at harvest (such as forage crops harvested to feed livestock) and can harbor insects and pathogens that damage crop plants.

Most weeds belong to one of two major groups, broadleaves (dicots) and grasses (monocots). Dicots start out having two seedling leaves (cotyledons); leaves usually have netlike veins. They are generally leafy and herblike (herbaceous), but they can also be woody plants (shrub- or treelike). Monocots produce only a single grasslike leaf as seedlings. Leaves of these plants typically have veins that run parallel to their length. Grasses, sedges, and rushes are monocots. Other less common weed types you may encounter include bryophytes (mosses and liverworts) and algae.

Weeds can also be divided into three groups according to life cycle: annual (summer or winter), biennial, and perennial.

Annual weeds live 1 year or less. They sprout from seeds, mature, and produce seeds for the next generation during this period. Annual weeds are either summer annuals or winter annuals (Fig. 2-2). Seeds of summer annuals sprout in the spring, and the plants produce seeds and die during the summer or fall. Some common summer annual weeds are little mallow, puncturevine, barnyardgrass, Russian thistle, and yellow foxtail. Seeds of winter annuals sprout in the fall and grow over the winter. These plants produce seeds in the spring and usually die before summer. Horseweed, wild oat, annual bluegrass, burclover, and Italian ryegrass are examples of winter annual weeds. When environmental conditions in an area are right for annual weeds, both winter and summer annuals may be found year-round.

Biennial weeds live for two growing seasons. They sprout and grow during the first season, then flower, produce seeds, and die the following season. Bristly oxtongue, poison hemlock, wild carrot, mullein, and Scotch thistle are biennial weeds.

Perennial weeds live 2 or more years; some species live indefinitely. Many perennials lose their leaves or die back entirely during the winter. These plants regrow each spring from roots or underground storage organs such as tubers, bulbs, or rhizomes. These storage organs are also the way many perennial weeds spread. Examples of perennial weeds are curly dock, silverleaf nightshade, field bindweed, yellow nutsedge, johnsongrass, and bermudagrass. Woody plants such as trees and shrubs are perennials and can be weeds when they are unwanted. Perennial weeds are the hardest type of weed to control.

The following sections describe some of the weeds you might find and wish to control on the land where you work.

FIGURE 2-2.

This illustration shows the difference in the growth periods of winter and summer annual weeds and the growth stages of perennial weeds.

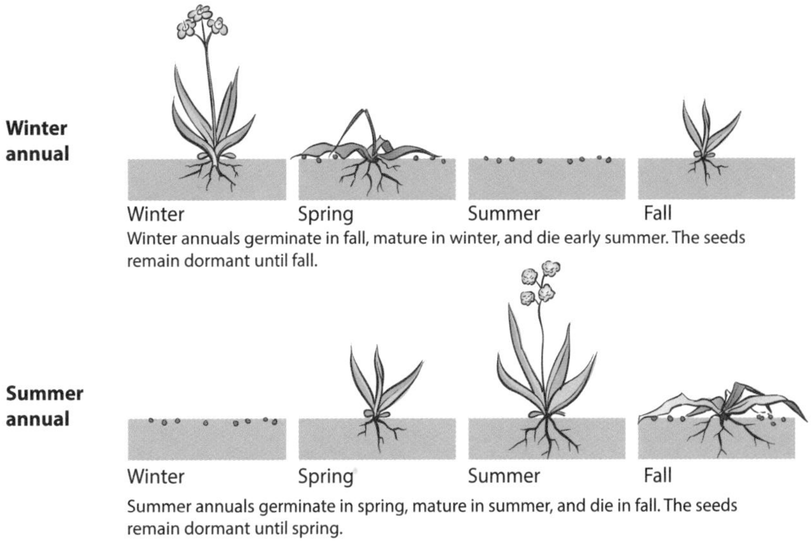

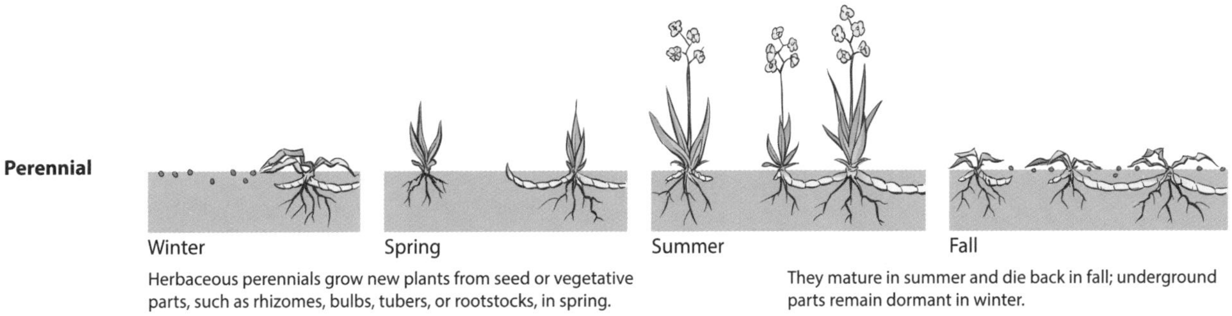

Perennial
Winter
Spring
Summer
Fall
Winter annuals germinate in fall, mature in winter, and die early summer. The seeds remain dormant until fall.
Herbaceous perennials grow new plants from seed or vegetative parts, such as rhizomes, bulbs, tubers, or rootstocks, in spring.
They mature in summer and die back in fall; underground parts remain dormant in winter.

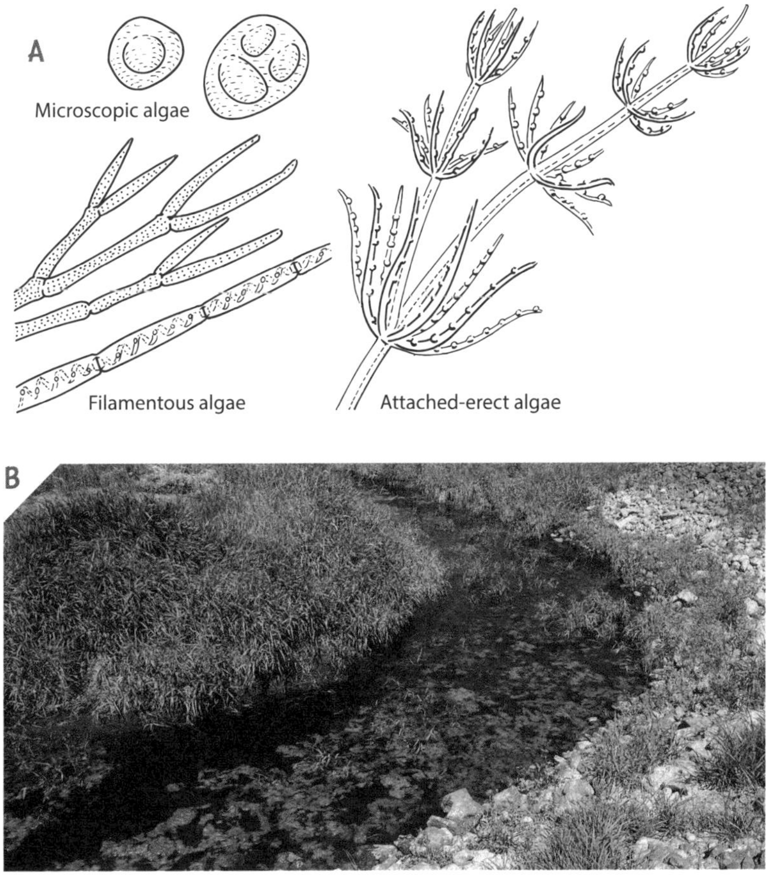

Algae

Important Characteristics. Algae are primitive plants closely related to some fungi. They reproduce by means of spores, cell division, or fragmentation. Pest algae fall within three general groups: microscopic, filamentous, and attached-erect (Fig. 2-3A). Microscopic algae impart greenish or reddish colors to water. They often float on the water surface as scums. Filamentous algae form dense, free-floating mats or mats attached to aquatic plants or rocks. Attached-erect algae look like flowering plants with leaflike and stemlike structures.

Algae clog irrigation channels, irrigation equipment, waterways, and ponds (Fig. 2-3B). Large algal buildup may use up the oxygen

FIGURE 2-3.

(A) There are three types of algae: microscopic, filamentous, and attached-erect. (Drawings are greatly enlarged.)

(B) Algae often clog waterways, as seen in this photograph.

within a body of water and kill fish. Some forms release toxins into water as they decompose, which may poison people or livestock.

Where Found. Algae occur in ponds, lakes, streams, rivers, and other bodies of water. Some forms of algae are problems in flooded rice fields and greenhouses.

Sedges and Rushes

Important Characteristics. Many sedges and rushes are perennial plants. They are grasslike and have fibrous root systems, and their perennial species produce rhizomes or tubers. Sedges (Fig. 2-4) have elongated, V-shaped leaves arising from triangular stems. Rushes have round leaves and a solid, round stem. These stem characteristics distinguish sedges and rushes from grasses, which have a hollow, round stem.

Where Found. Sedges are pests in orchards, vineyards, and irrigated crops, such as corn. They cause severe problems in rice fields. Sedges usually occur in marshy or poorly drained areas and along edges of ditches and ponds. Rushes are generally problems in aquatic systems and inhabit ecosystems similar to those of sedges.

Examples. Yellow nutsedge, purple nutsedge, blunt spikerush, and river bulrush.

FIGURE 2-4.

Sedges are grasslike plants with fibrous root systems. They are usually found in marshy or poorly drained areas.

FIGURE 2-5.

Grasses, such as this bermudagrass, spread quickly via stolons. The wide variety of ways grasses reproduce makes them difficult to control.

Grasses

Important Characteristics. Grasses (Fig. 2-5) are a large family of annual or perennial plants. They include many notable weeds as well as important crops, such as grains. Some species are the main food sources for grazing livestock. Roots of grasses are dense and fibrous; several species reproduce by rhizomes (underground stems). Many grasses can be identified by features of their collar region, as shown in Figure 2-6. Several important grassy weeds are winter annuals. Wild oat, for example,

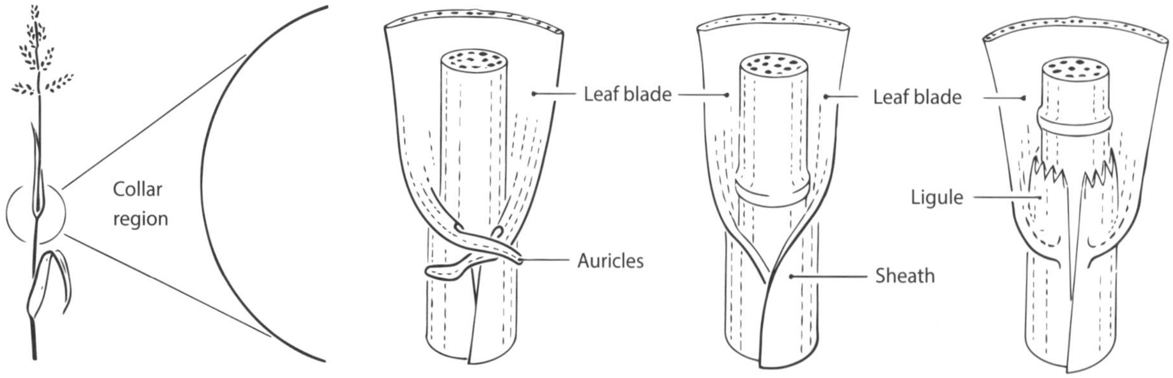

FIGURE 2-6.

The collar region of a grass leaf contains unique structures that are very important in identifying grass species.

is one of the most widely distributed, troublesome winter annual weeds in California.

Where Found. Most cultivated and natural areas contain grassy weeds. They are often pests in fields, pastures, rangelands, orchards, vineyards, along farm roads and ditch banks, and in other locations. Examples of affected crops include walnuts, prunes, grapes, kiwifruit, row crops, and most others.

Examples. Wild oat, bermudagrass, barnyardgrass, dallisgrass, annual bluegrass, yellow foxtail, and johnsongrass.

Broadleaves

Important Characteristics. Broadleaf weeds are plants with wide leaves that have many veins branching from a single main vein. They can be annuals, biennials, or perennials, and many can be identified as seedlings by their distinctive cotyledons. Some broadleaf weeds grow upright, like redroot pigweed (Fig. 2-7), and some grow low to the ground in a kind of mat, like field bindweed (Fig. 2-8). In cross-section, stems can be either round or angled (almost square). Some broadleaf weeds have specialized stems called rhizomes or stolons (stems that spread out on or near the surface) by which they reproduce (Fig. 2-9). Broadleaf roots can be fibrous, have a main taproot with smaller lateral roots, or

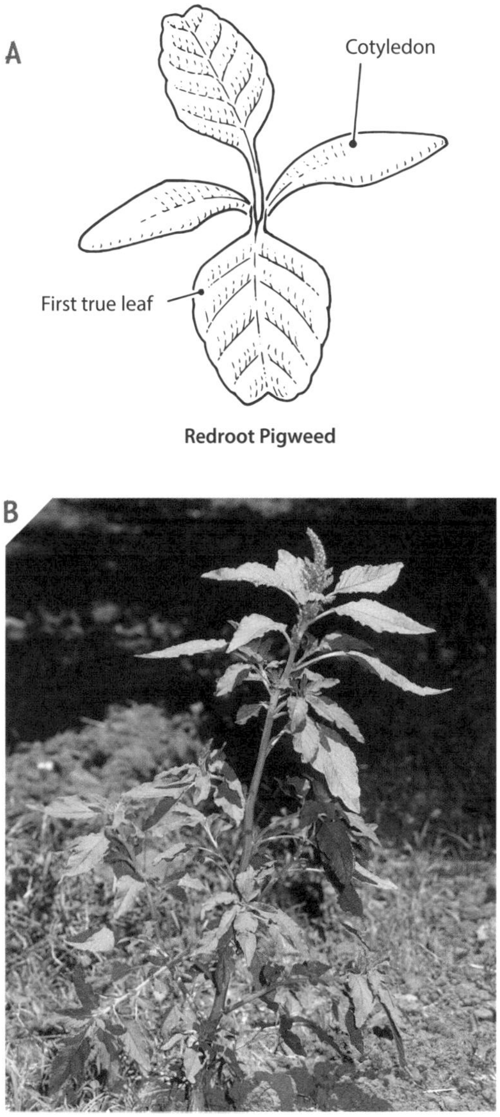

FIGURE 2-7.

Amaranthus retroflexus, amaranth family.

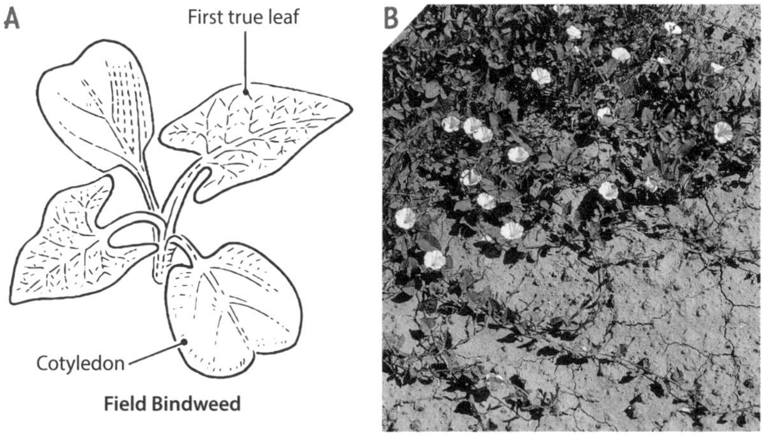

FIGURE 2-8.

Convolvulus arvensis, morningglory family.

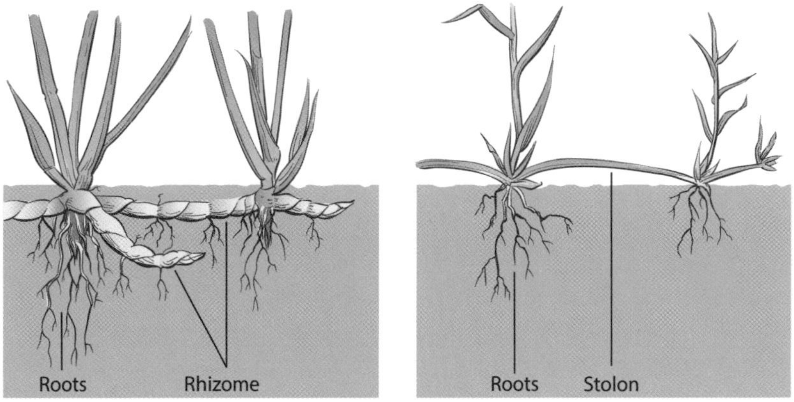

FIGURE 2-9.

Many weeds reproduce by underground and aboveground rooting structures such as those illustrated here.

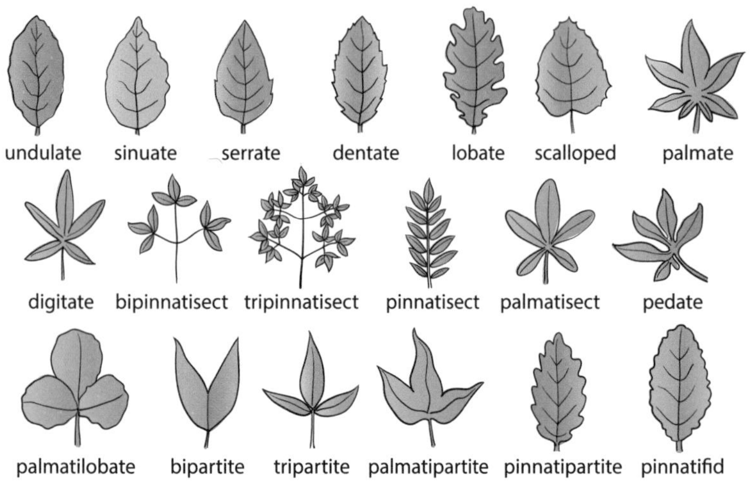

FIGURE 2-10.

The edges, or margins, of a leaf are often used to identify plants. *Source:* The Tree Center.

can be a hybrid of both types. Their leaves are arranged along a stem in specific ways, according to how each leaf attaches to the stem joint, or node. Leaves also have unique shapes, edges, and vein patterns, which can help you distinguish between weed species and crops (Fig. 2-10).

Where Found. Broadleaf weeds grow in almost all cultivated and undisturbed areas. These include rangelands and pastures, the edges of farm roads, ditch banks and fencerows, agronomic, horticultural, and vegetable crops, orchards, and vineyards. Examples of affected crops include grapes, prunes, almonds, walnuts, row crops, and most others.

Examples. Puncturevine, little mallow, redroot pigweed, field bindweed, jimsonweed, yellow starthistle, fleabane, and Russian thistle.

INVERTEBRATES

Invertebrates are animals without backbones (vertebrae). These include nematodes and all other microscopic worms, snails and slugs, and arthropods (insects, ticks, mites, and their relatives). Pest invertebrates affect people in many ways. Some are parasites of livestock or poultry; they feed on skin, hair, and blood or invade internal tissues. Many invertebrates transmit disease organisms to livestock, poultry, or plants. A large number of invertebrate pests are herbivores and feed on growing plants. Invertebrates also consume or contaminate stored commodities, and they can damage structures and equipment on the farm where you work. Table 2-2 lists some ways that invertebrates are pests in agricultural settings.

TABLE 2-2:

Ways in which invertebrates are pests

Type of pest	Type of damage	Example of invertebrate pest
plant pests	chewing on leaves	caterpillars, beetles, grasshoppers, snails/slugs
	boring or tunneling into leaves, stems, or fruit	twig borers, leafminers, beetles
	sucking plant juices	aphids, mites, scales, thrips, plant bugs
	feeding on roots	beetles, aphids, flies, nematodes
	feeding on fruits, nuts, berries	moth larvae, beetles, earwigs, snails/slugs
	causing malformations such as galls	flies, wasps, mites, nematodes
	transmitting diseases	aphids, mites, leafhoppers, nematodes
pests of animals	have venomous bite or sting	bees, wasps, ants, spiders, scorpions
	feed on flesh or blood	flies, mosquitoes, bugs, ticks, fleas, lice, mites
	transmit diseases	mosquitoes, bugs, flies, fleas, ticks
	cause loss in livestock weight gain or reduction of milk or egg production	fleas, ticks, mites
	damage and devalue hides and pelts; cause loss of carcasses used for meat	sheep ked, lice, ticks, mites, cattle grubs
	cause reduction in livestock's worth and reproduction efficiency	flies, mites, ticks

Ticks, Mites

Important Characteristics. Mites and ticks have their abdomen broadly joined to the head and thorax (Fig. 2-11). Adults usually have four pairs of legs, while immatures most often have three or fewer pairs. Some species of mites make fine webbing from silk glands that are near their mouths. Most mites are very small and hard to see without the aid of a hand lens or microscope.

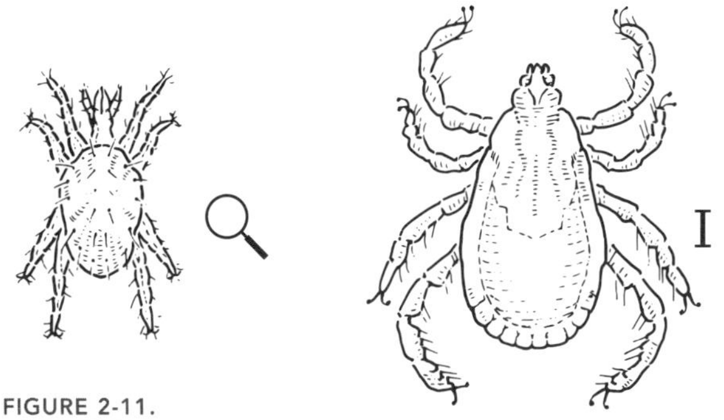

FIGURE 2-11.

Mites (left) and ticks (right) are closely related, although ticks are much larger. Ticks are parasites of vertebrates and feed on blood. Some mites are parasites of vertebrates, although many are serious plant pests.

Life Cycle. Ticks and mites hatch from eggs and pass through several immature stages before becoming adults. Immature ticks and mites resemble adults. Mites usually develop quickly from eggs to adults; some overwinter as adults, while other species overwinter as eggs. Ticks generally live much longer than mites. Some require 1 to 2 years to reach maturity and may live an additional 2 or 3 years as adults.

Where Found. Depending on the species, mites are parasites on plants or animals. Certain species are predatory on other mites. Plant-feeding mites can be found on upper or lower leaf surfaces. Ticks are blood-feeding parasites of vertebrates and require blood meals to develop and reproduce. They are commonly found on animal hosts or in or near their living spaces or nests.

Damage. Plant-feeding mites often produce serious economic damage, including leaf discoloration and defoliation. Some plant-feeding mites transmit pathogens (such as citrus leprosis virus). When mites feed on animals, their bites may itch severely. Toxins injected by ticks during feeding sometimes cause paralysis of hosts; some tick species transmit pathogens that cause disease (such as Lyme disease).

Beneficial Aspects. Several species of mites are predators of pest mites or small insects and are an important part of biological control programs.

Chewing Lice

Important Characteristics. Chewing lice (Fig. 2-12) are very small, oval or elongated wingless insects with chewing mouthparts. They have flattened bodies, sometimes with dark brown or black spots or bands. Chewing lice have a head that is wider than their thorax. You need a hand lens or microscope to really see these tiny insects.

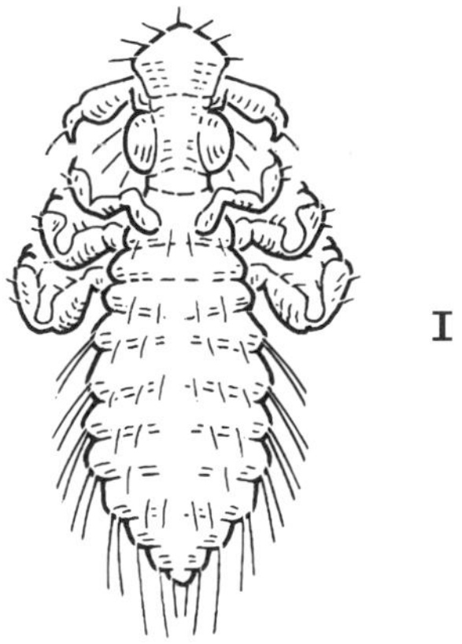

Chicken Head Louse

FIGURE 2-12.

Chewing lice, order Mallophaga.

Life Cycle. Chewing lice lay their eggs on hosts, usually attached to hair or feathers. They may pass through three or more nymphal stages before becoming adults. Most chewing lice develop into adults within 2 or 3 weeks after hatching.

Where Found. Chewing lice are parasites of birds, fowl, and livestock. Species are host-specific and feed on only one type of animal.

Damage. These parasites feed on feathers and the outer skin and skin debris of birds, and on the hair, blood, and skin of mammals. Poultry infested with chewing lice usually become restless and uncomfortable, have lower weight gain, and lay fewer eggs. On sheep, louse feeding can cause damage to fleece and wool.

Sucking Lice

Important Characteristics. Sucking lice (Fig. 2-13) are flat-bodied, wingless insects with piercing-sucking mouthparts. The head is narrower than the thorax. You need a hand lens or microscope to really see these insects.

Life Cycle. Females cement their eggs to the hairs of host animals. After hatching, sucking lice pass through several growth stages and become adults within 1 to 2 weeks. Sucking lice pierce the skin to feed on their host's blood.

Where Found. Sucking lice are host-specific parasites of mammals and typically feed on cattle and goats.

Damage. Feeding by sucking lice causes irritation and itching, anemia, reduction in milk production, and hair loss. Some sucking lice are capable of transmitting pathogens.

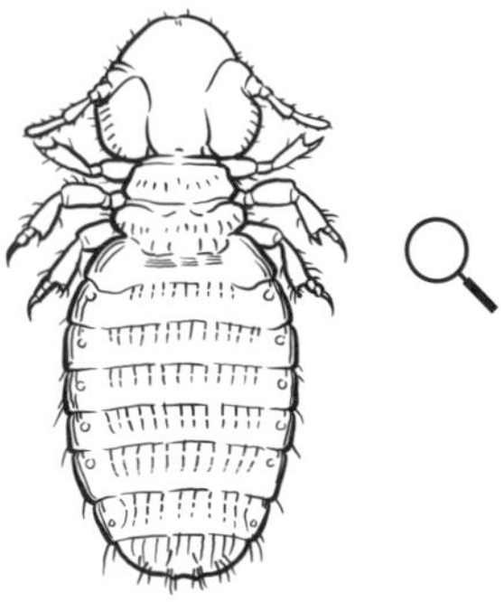

Cattle-Biting Louse

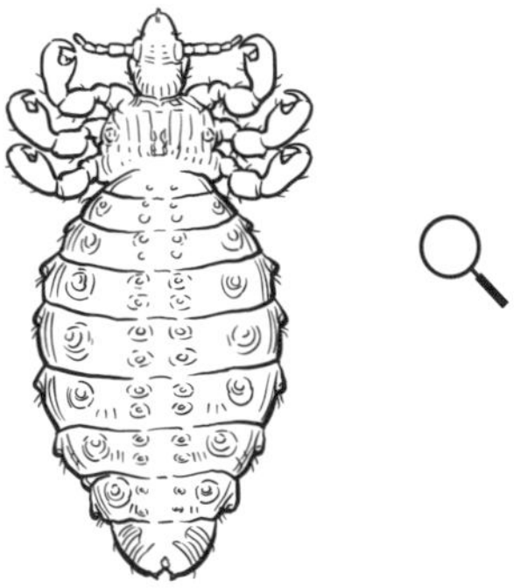

Shortnosed Cattle Louse

FIGURE 2-13.

Sucking lice, order Anoplura.

Thrips

Important Characteristics. Thrips are tiny, elongated insects with two pairs of wings (Fig. 2-14). Their wings have a kind of fringe. Thrips have modified sucking-rasping mouthparts.

Life Cycle. Thrips hatch from eggs, and most species pass through four growth stages. Thrips actively feed during their first two growth stages. The final growth stages are more of a resting stage, which often takes place in the soil. Wings are present on the adult thrips after the final molt.

Where Found. Thrips commonly infest plants and are often found in flowers and on tender, developing parts of leaves and fruits.

Damage. Thrips puncture plant cells and suck the fluid that leaks out. This type of feeding causes fruit and other plant parts to become deformed. Thrips feeding also creates black, wet spots on leaves that can only be seen under a microscope. Thrips can be serious pests in greenhouses, gardens, and agricultural areas, and are known to damage strawberry and citrus crops.

Beneficial Aspects. A few species of thrips are predatory. Predatory thrips play an important role in the natural control of several plant pests, including aphids and mites.

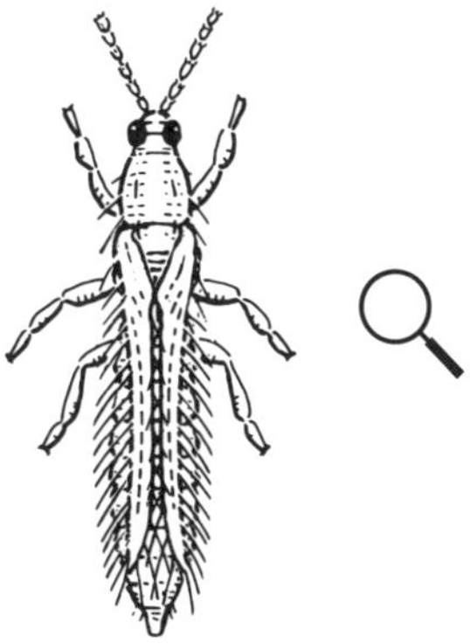

Western Flower Thrips

FIGURE 2-14.

Thrips, order Thysanoptera.

True Bugs (Chinch Bugs, Leaffooted Bugs, Lygus Bugs, Bagrada Bugs)

Important Characteristics. You can recognize most true bugs (Fig. 2-15) by the triangular-shaped plate or scale on the thorax seen from above. They also have a long, needlelike beak (piercing-sucking mouthparts) that folds under their bodies. Their forewings are hard at the base and thinner and more flexible at the tip. True bugs can be nearly

Consperse Stink Bug

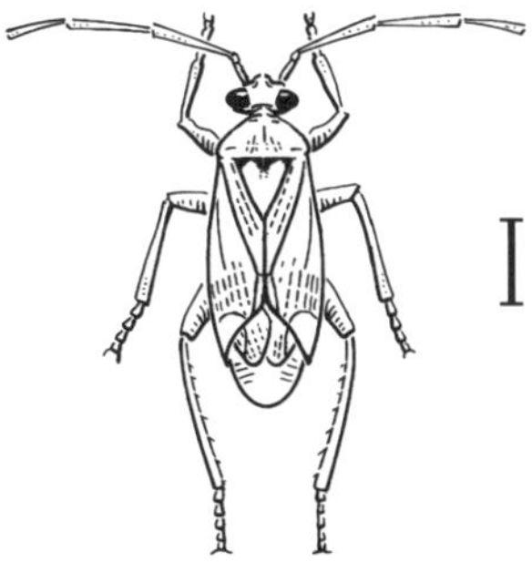

Lygus Bug

FIGURE 2-15.

True bugs, order Hemiptera.

2 inches long, but most are smaller than that. They have two pairs of wings and most fly well. The second pair of wings is only visible while the insect is flying. They sometimes have brightly colored wings.

Life Cycle. True bugs undergo incomplete metamorphosis after hatching from eggs. Young resemble adults but lack wings. Life cycles vary among the many species of true bugs.

Where Found. True bugs feed on plants and animals, depending on the species. Most are free living, searching out appropriate hosts for food. There are several aquatic species.

Damage. Plant-feeding true bugs damage plant cells. This causes deformities of fruits and other plant parts. Some true bugs also inject chemicals into the plant that prevent or alter the plant's normal growth. True bugs are often serious plant pests in agricultural settings. They are especially troublesome in cole crops, such as cabbage, kale, broccoli, and cauliflower, as well as in strawberries. There are some species that feed on the blood of livestock. Feeding sites may become inflamed or infected and usually are very tender, and a few species of true bugs transmit disease-causing organisms.

Beneficial Aspects. Some species of true bugs are predators of other insects, including many insect pests. Examples of these are assassin bugs, bigeyed bugs, and minute pirate bugs.

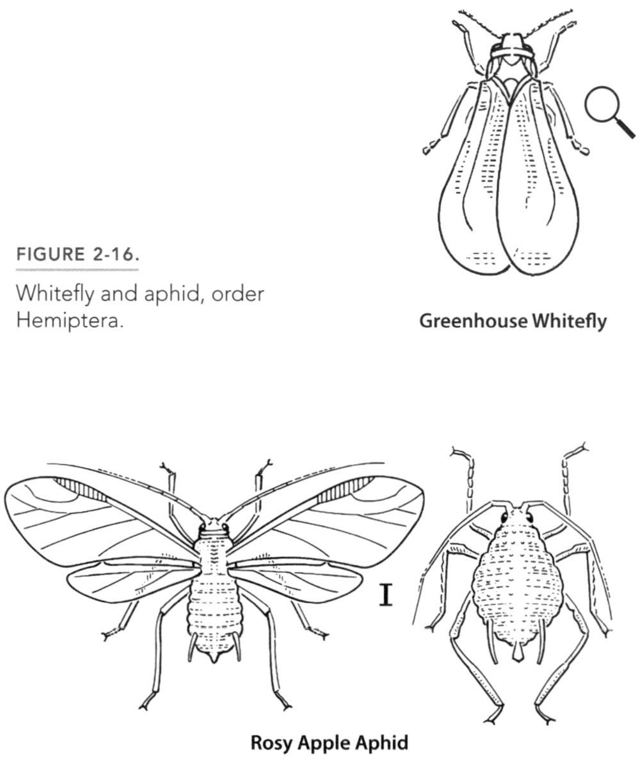

FIGURE 2-16.

Whitefly and aphid, order Hemiptera.

Aphids, Whiteflies, Mealybugs, Scales

Important Characteristics. These insects are diverse, somewhat soft-bodied, and most have wings or some winged forms. They all have piercing-sucking mouthparts. Figure 2-16 shows two of the many types of insects in this group.

Life Cycle. All insects in this group undergo incomplete metamorphosis, though the time it takes to go from egg to adult varies widely among species.

Where Found. These insects are plant feeders, so they are usually found on or near plants. Some occur in greenhouses.

Damage. Insects in this group pierce plant tissues and suck out liquids. Feeding usually causes deformed leaves and fruits, loss of plant vigor, stunted growth, and dieback of plant parts. Most of these insects excrete a sticky substance called honeydew that supports the growth of black sooty mold fungi. Many species transmit disease-causing pathogens to host plants. Aphids can be problems on prunes, pecans, and various vegetable crops; mealybugs damage pomegranates and grapes; scales damage almonds, walnuts, citrus, and stone fruits; and whiteflies are problematic on strawberries and most vegetable crops.

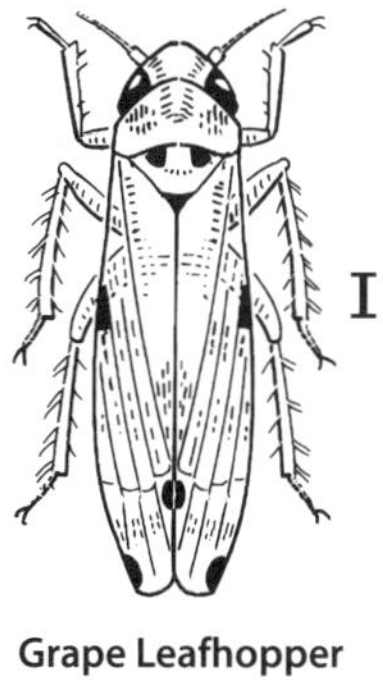

FIGURE 2-17.

Leafhopper, order Hemiptera.

Leafhoppers, Cicadas, Psyllids, Phylloxerans, Sharpshooters, Treehoppers

Important Characteristics. Almost all insects in this group have piercing-sucking mouthparts. Most of them feed on plant sap, but a few planthoppers feed on fungi or mosses. They have short, bristlelike antennae, and some mature insects develop wings (Fig. 2-17).

Life Cycle. All insects in this group undergo incomplete metamorphosis, though the time it takes to go from egg to adult varies widely among species.

Where Found. These insects can be found in many different habitats, including agricultural areas and nurseries.

Damage. Insects in this group feed on plants using piercing-sucking mouthparts, and many are vectors of viral and fungal diseases of plants. Their feeding causes scarring on fruits and can discolor or blacken leaves. The honeydew they leave behind is sticky, unsightly, and can result in the development of black sooty mold on infested plants. These insects are particularly damaging in grapes, citrus and other fruit trees, and vegetable crops.

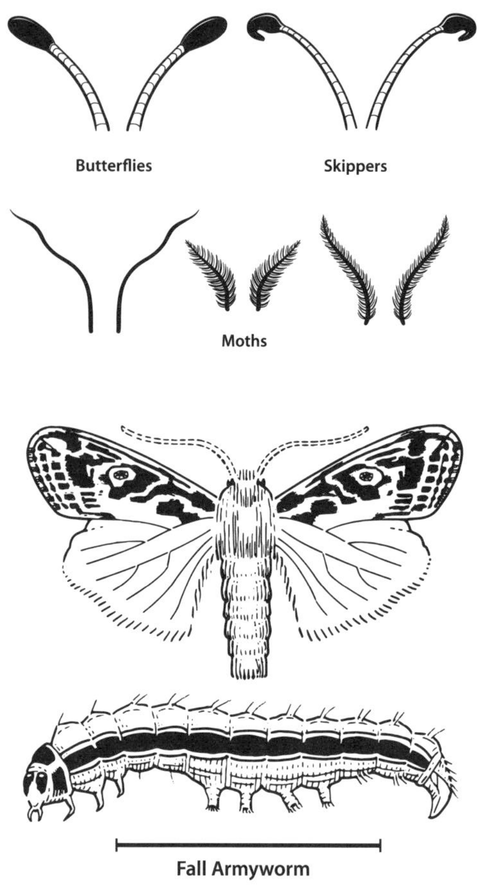

FIGURE 2-18.

Butterflies, moths, and skippers, order Lepidoptera.

Butterflies, Moths, Skippers

Important Characteristics. Adult butterflies, moths, and skippers (Fig. 2-18) have large, scale-covered, and often brightly colored wings that are different from the wings of most other insects. Larvae are wormlike, with chewing mouthparts. Adults have modified mouthparts in the form of a coiled tube that they use to suck up liquids. Butterflies and skippers can be distinguished from moths by their antennae, in most cases (certain rare moths and butterflies have antennae that resemble one another's). Skippers have shorter, thicker bodies and smaller wings, setting them apart from butterflies. Moths fly mostly at night, while butterflies and skippers fly mostly during the day.

Life Cycle. These insects undergo *complete metamorphosis*. After hatching from eggs, larvae pass through several growth stages, then enter the pupal stage and change into winged adults. A chrysalid or pupa can sometimes be found in the soil. Many of these pests overwinter in pupa form. Life cycles from egg through adult vary according to the species. Many species in this group produce three or four generations per year.

Where Found. These pests occur on or in plant parts (including fruits) and in stored food products. Adult moths are commonly attracted to lights.

Damage. Moth larvae, such as armyworm and codling moth, are one of the worst agricultural pests. They cause serious damage to fruits and vegetables, nuts, grains, cotton, and forage crops.

Flies, Mosquitoes, Gnats, Midges (Spotted Wing Drosophila, Leafminers, Blow Flies, Biting Stable Flies)

Important Characteristics. Adults in this group (Fig. 2-19) have only one pair of wings. In place of the second pair of wings are small clublike organs believed to assist in balance. Their larvae, known as maggots, are usually wormlike. Most adults have modified mouthparts for sucking, lapping, or piercing. Some adults have biting mouthparts.

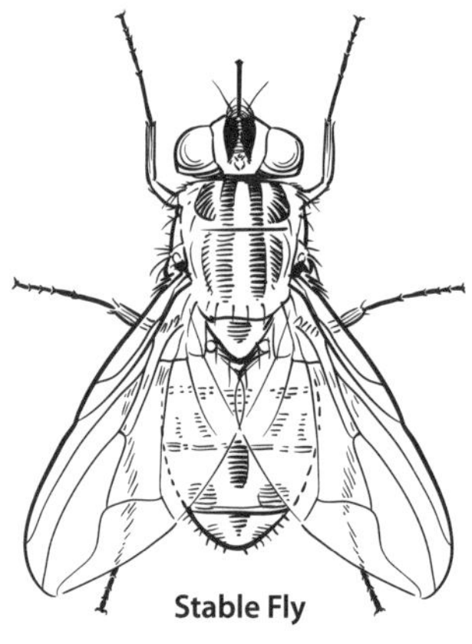

FIGURE 2-19.

Stable fly, order Diptera.

Life Cycle. These pests undergo complete metamorphosis. Most species lay eggs on surfaces or into tissues of hosts. In a few species, the eggs hatch inside the female's body, and larvae are deposited instead of eggs. Many species develop rapidly from eggs through the adult stages. This development may take as little as 3 or 4 days. Others have longer life cycles, taking 2 or more years to complete. Many species survive as resting pupae in the soil when conditions are not favorable for growth.

Where Found. Flies, mosquitoes, gnats, and midges occur in most outdoor areas and in areas where livestock and poultry are kept. Some larvae are internal parasites of animals; others invade plant tissues.

Damage. Many species in this group are serious pests. Larvae of some invade living animal tissues. Adult mosquitoes and the adults of some species of flies, midges, and gnats feed on blood. Many of these insects transmit serious disease-causing pathogens. Some species of flies are pests in agricultural crops, such as caneberries, strawberries, cole crops (cabbage, kale, broccoli, etc.), and grapes. Others cause problems in and around livestock or poultry operations.

Beneficial Aspects. A few fly species are parasitic on pest insects, and others are predators. These species are often an important part of biological control programs.

Snails and Slugs

Important Characteristics. Snails prefer cool, damp places. If they cannot find suitable shelter, their shells will protect them. They can seal themselves into their shells and become dormant for up to 4 years during dry periods. Slugs also prefer cool, damp areas but do not have protective shells and so are vulnerable to high temperatures and dry weather.

Life Cycle. Snails and slugs lay between 10 and 200 eggs under the soil. They lay egg masses several times each year from spring through fall. In cooler areas, egg laying stops during the winter. It takes from 1 to 3 years for newly hatched snails or slugs to reach maturity.

Where Found. Snails and slugs live in damp areas, soil litter, and plant foliage. They are usually active at night, hiding during the day under boards or stones, or among ivy, dense shrubbery, or damp litter. At night and early in the morning, or during cool, damp periods, they look for food. They often return to the same resting area each day unless it becomes too dry or disturbed.

Damage. Snails and slugs feed on foliage, fruits, berries, and vegetables. Snails can be serious pests in citrus, where they feed on developing fruits. They are also pests in greenhouses. Besides feeding damage, both snails and slugs leave slime trails that detract from the appearance of produce and foliage.

Beneficial Aspects. Some species of snails are predators, making them useful in biological control programs aimed at pest snails.

VERTEBRATES

Vertebrate pests include fish, amphibians (frogs, toads, and salamanders), reptiles (turtles, lizards, and snakes), birds, and mammals (ground squirrels, gophers, and coyotes). They become pests if they

- host pathogens that cause disease (such as plague and rabies)
- damage crops or stored products
- create conditions that favor other pests (like weeds)
- prey on livestock
- interfere with the activities or needs of people

FIGURE 2-20.

Birds can be pests when they damage agricultural crops, spread disease to poultry, or foul stored commodities.

Birds

Birds can become pests when they help spread pathogens that can be transmitted to poultry or when they eat or damage crops, such as grapes and stone fruits (Fig. 2-20). Some bird species can be removed only with a depredation permit from the California Department of Fish and Wildlife or while under the supervision of the local county agricultural commissioner. Check with your local Fish and Wildlife warden or agricultural commissioner before removing birds, even on your own property or property you manage.

Mammals

Pocket Gophers. Pocket gophers (Fig. 2-21) are burrowing rodents that get their name from the fur-lined external cheek pouches, or pockets, that they use for carrying food and nesting materials. They are well equipped for digging and tunneling with their powerful front legs, large-clawed front paws, and fine, short fur that doesn't cake in wet soils. Their small eyes and ears and highly sensitive facial whiskers assist them in moving around in the dark. Gophers don't hibernate and are active year-round, although you might not see any fresh mounding. They also can be active at all hours of the day.

Gophers live alone within their burrow system, except when females are breeding or caring for their young. You can find as many as 60 or more gophers per acre in irrigated alfalfa fields or in vineyards. Gophers reach sexual maturity at about 1 year of age and can live up to 3 years. In nonirrigated areas, breeding usually happens in late winter and early spring, resulting in one litter per year; in irrigated sites, gophers can have up to three litters per year. Litters usually average five or six young.

FIGURE 2-21.

Adult pocket gophers are rarely seen above ground except when pushing soil from their burrow, as shown here, and sometimes when clipping small plants near a burrow opening.

Pocket gophers often invade fields and orchards, feeding on the roots and trunks of many food and forage crops, vines, and trees. They are especially troublesome in citrus groves and alfalfa fields. A single gopher moving down a planted row can cause lots of damage in a very short time. Gophers also gnaw and damage buried irrigation systems. Their tunnels can divert and carry off irrigation water, which wastes water and leads to soil erosion (Salmon and Baldwin 2009).

Ground Squirrels. Ground squirrels (Fig. 2-22) are common pests of alfalfa and citrus crops, as well as in almond, apple, apricot, peach, pistachio, prune, and walnut groves. Ground squirrels live in colonies that may grow very large if left uncontrolled. They are active during cooler times on hot days as well as sunny periods during the cooler months; they are usually most active in morning and late afternoon. In periods of high winds, ground squirrels retreat to their burrows. This squirrel's habitat includes nearly all regions of California except for the Owens Valley.

Ground squirrels breed once a year, averaging seven or eight young per litter. Aboveground activity by adults is highest during the main part of the breeding season. The young are born in the burrow and grow very fast. Young squirrels usually come out of their burrows at about 6 weeks old. At 6 months they resemble adults.

FIGURE 2-22.

Ground squirrels are often pests because they compete with people for agricultural products. Squirrels also damage levees and bridge foundations through their burrowing activities. Some vector diseases that fleas or other insects transmit to people and livestock.

Ground squirrels damage young vines and trees when they gnaw bark, girdle trees, and eat twigs and leaves. Their burrowing around tree roots and in fields and other areas of a property can be very destructive.

Diseases and Disorders

Plant and animal diseases are caused by pathogens such as fungi, bacteria, viruses, viroids, and phytoplasmas. Disorders are caused by nonliving (abiotic) factors, such as nutrient deficiencies or pollution. Diseases and disorders alter or interfere with the chemical processes that take place within an organism's cells. To avoid unnecessary pesticide treatments, you must accurately identify the cause of the symptoms you see in plants and animals.

Fungi

Identification. Fungi are mostly microscopic organisms, but some have forms, such as mushrooms, that are large and can be seen without magnification. The body of the fungus is made up of tiny tubular filaments called hyphae. A mass of hyphae growing together is called the mycelium. Reproductive structures called spores may be produced on specialized hyphae or fruiting bodies.

Spores and spore-bearing structures are what people use to identify fungi. Size, shape, color, arrangement of spores, and the shape and color of the fruiting body can be used to help identify the fungus (Fig. 2-23).

Types of damage that may be caused by fungus infection include leaf spots; blight on leaves, branches, twigs, and flowers; cankers; dieback of twigs; root rot; seedling damping off; stem rots; soft and dry rots; scab on fruits, leaves, or tubers; and the overall decline of the host. All of these symptoms also contribute to the stunting of infected plants.

Fungal infections can also result in deformed plant parts. Examples of symptoms you can see include enlarged roots (clubroot); enlarged growths filled with mycelia (galls); warts on tubers and stems; profuse upward branching of twigs (witches'-broom); and distorted, curled leaves (leaf curl). Other symptoms include wilting or powdery buildup (rusts and mildews).

Field identification of disease symptoms caused by fungi can often help determine the species. However, many disease symptoms look alike, so you should collect samples to send to an expert for positive identification (for help finding identification resources, see "Identification Experts" earlier in the chapter).

Life Cycle. The life cycles of fungi vary. Generally, fungi reproduce via spores. Spores are specialized reproductive bodies that may be formed sexually or asexually. Almost all fungi have an asexual cycle, and asexual reproduction can occur several times within a growing season. In most of the fungi that have a sexual reproductive cycle, reproduction occurs only once a year.

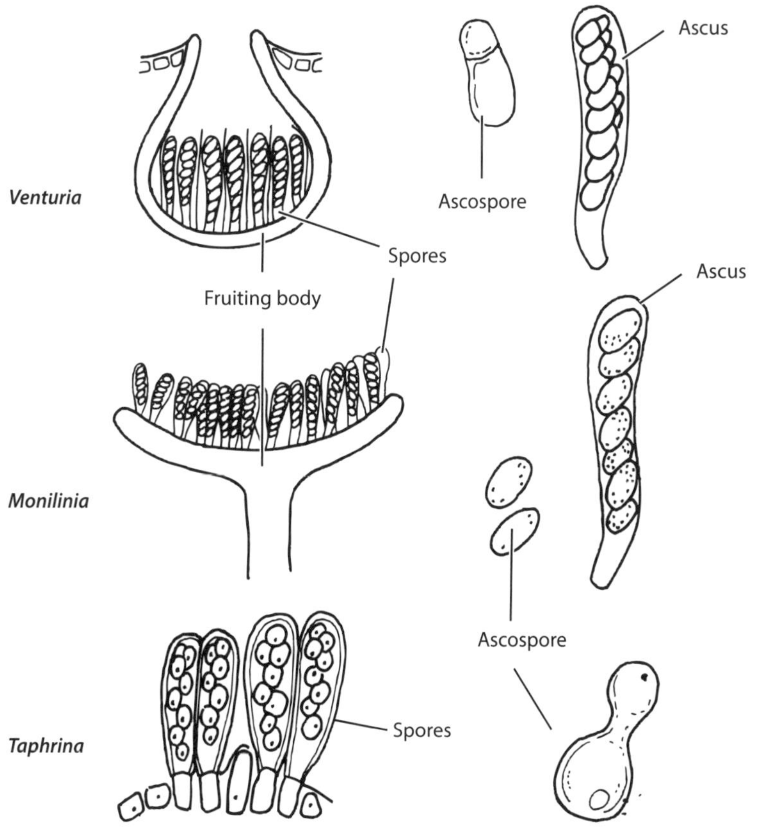

FIGURE 2-23.

These footprint-shaped spores were produced by *Venturia inaequalis*, which causes apple scab. These drawings show fruiting bodies, ascospores, and asci for the apple scab pathogen in the genus *Venturia* along with two other types: the genus *Monilinia*, which includes the agent causing brown rot of stone fruits, and the genus *Taphrina*, which includes the agent causing peach leaf curl.

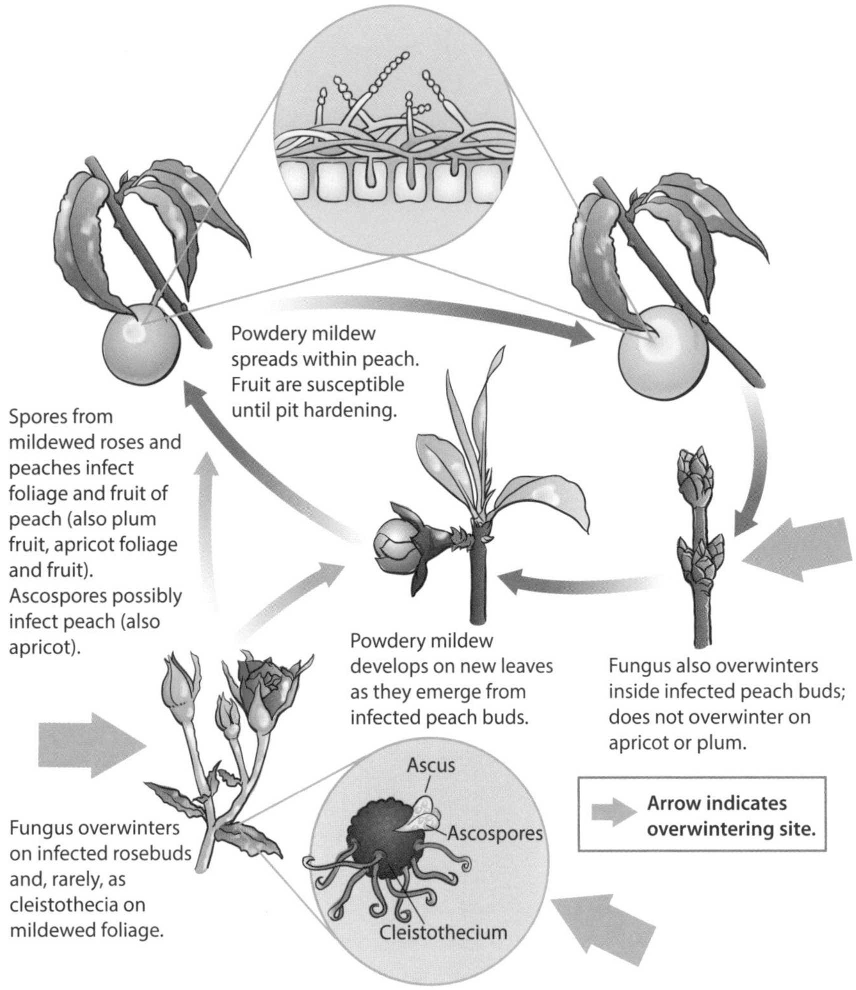

FIGURE 2-24.

Seasonal cycle of powdery mildew caused by the *Sphaerotheca pannosa* fungus in apricots, peaches, and plums. The pathogen can overwinter as mycelia on rosebuds or peach buds or as cleistothecia on rose foliage, but it cannot survive on plum or apricot.

Fungi can survive in different ways when conditions are unfavorable (Fig. 2-24). Fungi overwinter as mycelia or spores in or on infected tissue. Resting spores of some species are resistant to extremes in temperature and moisture. Infected plant litter, bud scales, or bark cankers in trees, shrubs, or soil can provide overwintering sites for the mycelia, sclerotia, or spores of many fungi. Seeds and other vegetative organs or alternate hosts such as weeds provide other survival sites for overwintering fungi.

Dispersal and Movement. A few fungi are able to move from one host to another by extending *rhizomorphs* through the soil; however, most fungal spores spread in air currents and can be carried over long distances. Spores, sclerotia, and mycelial fragments can be spread by water; some fungi need rain to spread. Insects can spread fungal pathogens. Animals, people, and dirty equipment can carry fungal spores. People also spread pathogens when they move infected seeds, transplants, nursery stock, and used containers.

Beneficial Aspects. Certain beneficial fungi form cooperative bonds with plant roots; work against soil, foliar, or fruit diseases; or attack pest insects, mites, and nematodes. These fungi can be an important part of IPM programs.

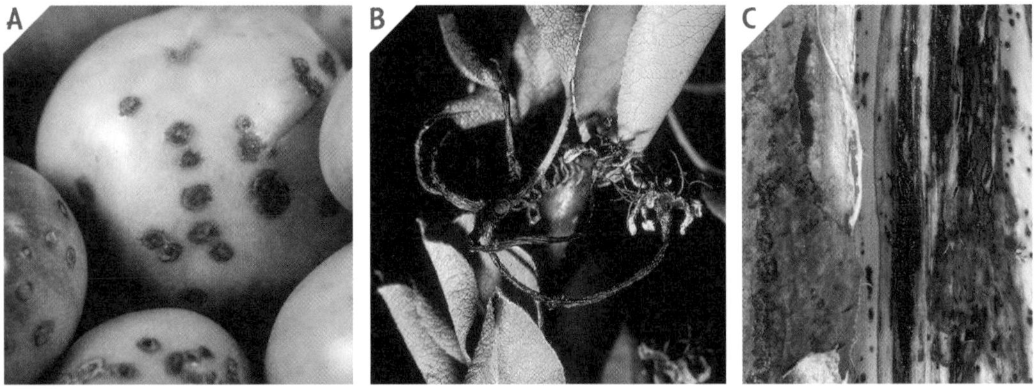

FIGURE 2-25.

Disease symptoms of some bacterial diseases. Bacterial spot lesions caused by *Xanthomonas vesicatoria* on tomato fruit have a rough, scabby texture (A). When they first develop, the lesions are surrounded by a white halo, similar to bacterial speck. As with all bacterial diseases, laboratory analysis is required for identification. Pear fruit, flowers, and leaves infected with fire blight turn black (B). Tan droplets of ooze are also typically present. A walnut trunk damaged by deep bark canker bacteria, *Erwinia rubrifaciens*, typically shows dark spots underneath the bark (C). *Source:* Flint 2012.

Bacteria

Identification. Bacteria are very small, typically under 0.002 mm (or about one twelve-thousandths of an inch) long. Field diagnosis of bacterial disease relies on recognition of disease symptoms (Fig. 2-25). Plants generally have

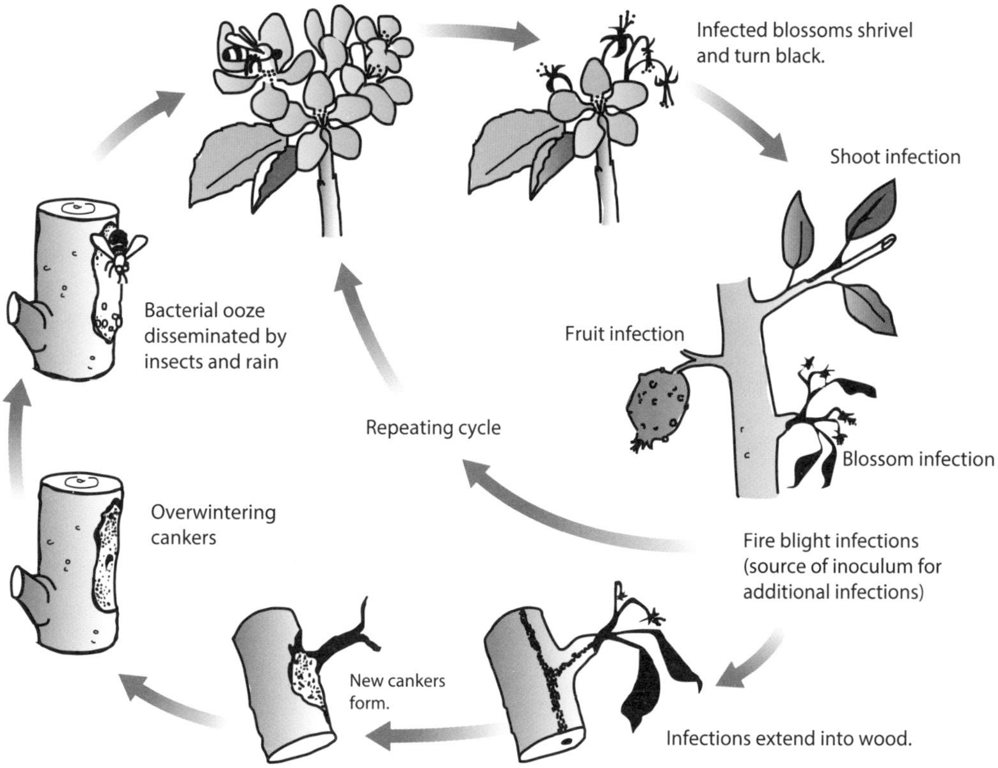

FIGURE 2-26.

The disease cycle of fire blight of pear and apple caused by the bacterium *Erwinia amylovora*. The bacteria overwinter at the margins of cankers on branches and twigs and are disseminated in spring when trees start growing and sap flows out of the cankers. Splashing rain and insects attracted to the ooze move the bacteria to blossoms. Bacteria multiply rapidly in the blossoms and enter other tree tissues through stomata, lenticels, and wounds.

specific reactions to bacterial infections; however, laboratory analysis is required for positive identification. Common symptoms include cankers, galls, wilts, slow growth, rots, discoloration of plant parts, deformed fruits or leaves, slow ripening, brooming, and leaf spots. Some bacterial diseases produce a slimy ooze on plant surfaces.

Certain bacteria cause galls to form on specific plant parts; examples include olive knot and crown gall. Galls may interfere with the movement of food and water in the plant. Bacterial wilts generally affect the entire plant; the infection produces slime that plugs the water-conducting tissue of the infected plant. Destructive wilts include those found in tomato, cotton, cucumber, and corn. The bacteria that cause cankers destroy plant tissues, as seen in plants with tomato canker and fire blight. You might see areas of spotting, most commonly on leaves and fruits; potato scab, for instance, appears as local spots on tubers. Rots invade fleshy tissue and often cause a slimy, foul-smelling ooze. Soft rots are often secondary infections that appear only after another pathogen has invaded.

Growth Requirements. Bacteria enter plants and animals through natural openings and wounds. Once inside, some bacteria move along the sap stream or bloodstream; others move about with the flow of fluids between the cells. In early stages of disease in plants, bacteria often develop between the cells. As cell walls are injured, the bacteria enter the cell and continue growing. The disease cycle of fire blight is shown in Figure 2-26.

Disease symptoms appear after bacteria enter and develop in an organism's tissue. For example, soft rots and leaf spots are usually visible within a day; crown gall infections can take up to 2 years to become evident. The time between infection and the appearance of symptoms is called the incubation period.

Survival Characteristics. Bacteria survive mostly within plant hosts as parasites, on seeds, or in plant litter in soil. They do not produce overwintering spores like fungi; however, many are able to overwinter in plant litter.

Dispersal and Movement. Bacteria can be carried from one place to another in splashing or flowing water and rainwater, by moisture, and during certain management activities. Bacteria that survive in organic matter in the soil can be spread by any activity that moves soil from one place to another, such as cultivation. Bacterial pathogens are often carried on infected seeds, cuttings, or transplants. Insects and animals can also aid in spreading bacteria. Cucumber beetles, for example, transmit cucumber wilt bacteria, and sharpshooters transmit the bacteria that cause Pierce's disease.

Viruses

Identification. Because of their small size and transparency in the host's cells, viruses cannot be seen or detected in the same way as other pathogens. Therefore, your main detection clues will be the plant's symptoms.

Viral diseases produce a variety of symptoms in plants. Common field symptoms of plant viruses include growth reductions, color changes, deformities, and necrosis (death) of tissue; severe stunting and reduced yields may be seen. Mosaic patterns (Fig. 2-27), a mottling of healthy and discolored tissue on leaves, are a common virus symptom. Some viruses roll or crinkle leaves; the affected leaves may be a deeper green. Vein clearing or leaf yellowing is typical of some viral infections. Some viral diseases cause the plant to become dwarfed, while others stimulate short, sporadic shoots or stunting and rosetting (Fig. 2-28). Positive identification of a virus in a plant can be made only by specialized laboratory tests.

FIGURE 2-27.

Discolored bands, lines, and ring patterns, such as on these leaves infected with alfalfa mosaic virus, are typical symptoms of certain viruses.

FIGURE 2-28.

Some viruses can cause severe stunting and tight, bunchy growth (rosetting). This almond tree has been infected with yellow bud mosaic for several years and shows severe stunting and concentration of leaves on terminals.

Growth Requirements. All viruses are parasitic on cells and require a host cell for survival and reproduction. They are not cellular organisms themselves.

Viruses enter plants through wounds made by insect or nematode *vectors*, mechanical injury, or from an infected pollen grain or grafting. The first virus particles may appear about 10 hours after a virus enters a plant *(inoculation)*. For infection to occur, the virus must move from cell to cell and multiply in those cells. The movement of viruses in the plant varies with the virus and the host.

Survival Characteristics. Viruses cannot survive in dead plant matter or outside living plant tissue. Alternate hosts such as perennial plants, weeds, or volunteer crop plants provide overwintering sites for many viruses and are often suitable hosts for virus vectors as well. Insect vectors also provide important overwintering sites for viruses.

Dispersal and Movement. Viruses can enter plants only through wounds. They are spread from plant to plant when you use cuttings to produce more plants *(vegetative propagation)*; through sap, seeds, and pollen; by vectors; and by the use of contaminated pruning tools. Vectors of one or more plant viruses include aphids, leafhoppers, whiteflies, beetles, thrips, mites, nematodes, fungi, and dodders. Vectors are specific to certain viruses and hosts, such as mealybugs that spread leafroll virus in grape and beet and leafhoppers that spread curly top virus in tomatoes.

Abiotic Factors Causing Disorders

Abiotic disorders are noninfectious diseases caused by bad environmental conditions, often as a result of human activity. Their symptoms can resemble damage caused by pests. Abiotic disorders can be caused by nutrient deficiencies or excesses; low or high temperatures; toxic levels of salt or pesticides; air pollution; and too little or too much water (Table 2-3). Activities that compact soil, change the soil grade, or injure trunks or roots can also result in diseased plants. In addition to direct damage, abiotic disorders can weaken plants that are then more easily attacked and damaged by insects and pathogens.

Although some abiotic disorders can be recognized by symptoms such as deformed or discolored foliage, roots, stems, fruits, or flowers, abiotic disorders are hard to identify with certainty. A field history, records of pesticide and fertilizer use, and tests on soil or leaf tissue samples may help in diagnosis. Patterns of diseased plants in the field can also help you identify abiotic disorders. In general, symptoms of abiotic disorders start suddenly and do not spread through a plant or to other plants over time as pest damage can.

TABLE 2-3:

Common abiotic disorder symptoms and their causes

Symptoms*	Possible cause
Foliage wilts, droops, discolors, and drops earlier than expected. Twigs and limbs may die back. Bark cracks and develops cankers. Plant may be attacked by wood-boring insects.	water deficiency
Foliage yellows and drops. Twigs and branches die back. Root crown diseases develop.	water excess or poor drainage
Foliage is discolored, undersized, sparse, or distorted and may drop earlier than expected. Leaves turn yellowish or brownish, especially along margins. Plant growth is slow. Limbs may die back. Bark becomes corky.	mineral deficiency
Foliage turns brown, dry, and crispy. Limbs may die back. May smell of natural gas in the area.	natural gas line leak underground
Foliage or shoots turn yellowish, undersized, or distorted. Leaves may appear burned, with dead margins, and may drop.	pesticide toxicity
Yellow, brown, then white areas develop on upper side of leaves or fruits. Foliage may die.	sunburn
Leaves or needles turn yellowish, brownish, or have discolored flecks. Foliage may be sparse, stunted, and drop earlier than expected.	air pollution
Foliage is unusually dark.	excess light
Excess growth of succulent foliage. Foliage may appear burned and die. Plant infested with many mites, aphids, psyllids, or other insects that suck plant juices.	nitrogen excess
Shoots, buds, or flowers curl, darken, and die. Limbs and entire plant may die.	frost
Foliage, twigs, or limbs injured. Cankers may develop.	hail or ice
Bark or wood dead, often in a streak or band.	lightning
Bark is cracked or sunken, often on south and west sides. Wood may be attacked by boring insects or decay fungus.	sunscald

Note: *Many of these symptoms can have other causes, including pathogens and insects.
Source: Flint 2012.

Chapter 2 Review Questions

1. Which of the following is *most* likely to happen if you identify pests incorrectly?

☐ a. Pests may escape before they can be killed by pesticide applications.
☐ b. You may confuse beneficial insects with pest insects in the field.
☐ c. Your pest control efforts will often fail regardless of site conditions.

2. Match the characteristics with the weed type.

1. leaves have parallel veins	a. monocots
2. plants are shrub- or treelike	
3. leaves have netlike veins	b. dicots
4. seedlings have a single leaf	

3. Invertebrate pests include which of the following organisms?

☐ a. crayfish, shrimp, and eels
☐ b. slugs, snails, and salamanders
☐ c. ticks, mites, and nematodes

4. Which organisms belong in each of the numbered pest groups?

1. vertebrates	a. ticks
	b. little mallow
2. invertebrates	c. ground squirrels
	d. powdery mildew
3. weeds	e. yellow nutsedge
	f. aphids
4. pathogens	g. fire blight
	h. pocket gophers

5. Only trained experts using special methods and equipment can positively identify pests such as ________ and ________.

☐ a. ticks, mites
☐ b. nematodes, pathogens
☐ c. gophers, rats

6. Symptoms of abiotic disorders start suddenly and ____________.

☐ a. do not spread through a plant or to other plants
☐ b. spread quickly from one plant to another
☐ c. are unlikely to kill the affected plant

7. Annual, biennial, and perennial are the three possible life cycles of _______.

☐ a. weeds
☐ b. vertebrates
☐ c. pathogens

8. Which of these pests attack both plants and animals on a farm?

☐ a. thrips
☐ b. mites
☐ c. lice

CHAPTER 3
PESTICIDES

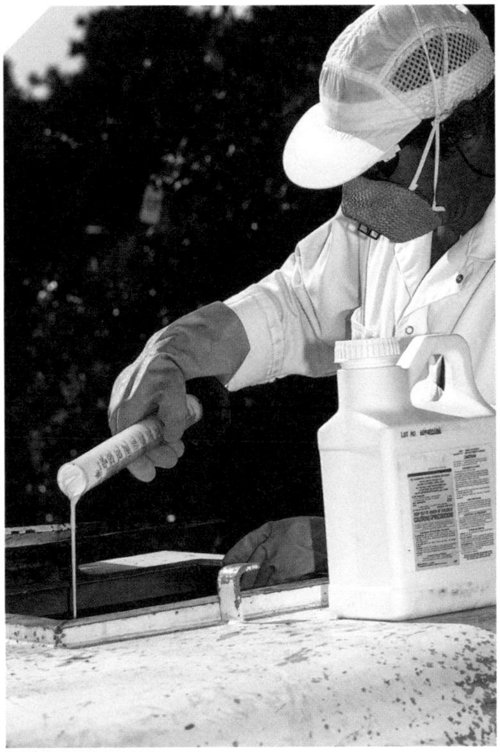

Pesticides and Their Hazards 30
How Pesticides Are Organized 32
Adjuvants 38
Chapter 3 Review Questions 39

Knowledge Expectations

1. Explain the concepts of hazard, exposure, and toxicity and how they relate to one another.
2. List pesticide toxicity categories and signal words, and explain what each category means in terms of a pesticide's effects on humans and animals.
3. Identify factors that should be considered when selecting pesticides.
4. List groups of pesticides according to pest target, and describe the functions of each group.
5. List major chemical families, and describe the particular hazards associated with each one.
6. Define mode of action, and provide examples of the different modes.
7. Explain how various modes of action influence pesticide selection.
8. Explain how contact and systemic pesticides control pests differently.
9. Explain why pesticides are sold as formulated products.
10. List the various formulations available and the advantages and disadvantages of each.
11. Explain the role of adjuvants in pesticide applications.

Pesticides and Their Hazards

A pesticide is any substance that is used to control, prevent, eradicate, or repel any type of pest organism. For instance, adjuvants that are added to a tank mix to improve deposition are considered pesticides because they are used as part of a program to control or eradicate pests. Any product used to defoliate or regulate the growth of a plant for any reason is also considered a pesticide. California regulations have a specific definition of the term, which can be found in Appendix A.

PESTICIDE HAZARD, EXPOSURE, AND TOXICITY

Explain the concepts of hazard, exposure, and toxicity and how they relate to one another.

All pesticides are toxic. They must be toxic to kill the pests you are trying to control. Some pesticides are more toxic than others. The hazard to you and others when you use pesticides is a combination of this toxicity and the amount of exposure. Exposure can take place through several routes–your skin, eyes, mouth, and lungs–and the route of exposure may influence the degree of hazard.

A number of factors affect the toxicity of a pesticide as it is being used and after it has been applied. These include the passage of time, characteristics of the water used for mixing, features of the application site (soil type, location, etc.), formulation and dosing, and chemical reactions that occur during mixing. Once applied, pesticides usually break down into different chemicals or chemical compounds over time. These new chemicals may be less toxic or more toxic than the original pesticide.

The time it takes for half of what you apply to break down into its component parts is called the pesticide's half-life. Soil microbes, ultraviolet light, temperature, quality of the water used in mixing, or impurities combined with the pesticide increase or decrease the half-life and can influence toxicity. Sometimes impurities contaminate pesticides during manufacture, formulation, storage, or while you are mixing them. In addition to the environment, the chemical nature of the pesticide, its formulation, and the dose applied affect its toxicity. Mixing two or more pesticides together can also change their toxicity or alter their half-life.

PLANT AND ANIMAL TESTING

One way to measure the toxicity of pesticides is to give known doses to laboratory animals and observe the results. Animal testing is the way researchers find out the lethal dose or lethal concentration of each pesticide. Through animal testing, researchers also decide the maximum dose to which organisms can be exposed without causing injury. They use the results from these types of tests to predict hazards to people and nontarget organisms.

Research workers test pesticides on mice, rats, rabbits, and dogs. They also perform toxicity tests on nontarget plants and animals if these organisms are at risk from pesticide exposure. Nontarget animals may include insects (such as bees), fish, amphibians (frogs, toads, salamanders), deer, birds, and other wildlife. Researchers also test pesticides on target pests to set the dosage rates listed on pesticide labels. These tests also tell how well the pesticide works under different conditions. The effectiveness of a pesticide on its target pest is known as its efficacy.

Lethal Dose and Lethal Concentration. Researchers divide laboratory animals into several groups and test different routes of exposure (skin, mouth, eyes, lungs). They rate a pesticide's toxicity by determining the amount that kills 50% of a test population. This level is the lethal dose, or LD_{50}, which is expressed as the milligrams of pesticide per kilogram of body weight of the test animal (mg/kg). Researchers also determine how much pesticide vapor or dust in the air or what amount of pesticide diluted in rivers, streams, or lake water causes death in 50% of test animal populations. This level is the lethal concentration, or LC_{50}, which is expressed as micrograms (1 one-millionth g) per liter of air or water (µg/l).

PESTICIDE TOXICITY CATEGORIES

List pesticide toxicity categories and signal words, and explain what each category means in terms of a pesticide's effects on humans and animals.

Federal regulations place pesticides into one of four categories according to their toxicity and potential to injure people, animals, or the environment (Table 3-1). Pesticide labels indicate these categories by the following signal words:

- Category I, DANGER or DANGER-POISON
- Category II, WARNING
- Category III, CAUTION
- Category IV, CAUTION or no signal word

TABLE 3-1:

Signal words of pesticide toxicity categories that appear on U.S. Environmental Protection Agency (EPA)–approved labels

	Pesticide label signal words		
Hazard indicators	**DANGER/DANGER-POISON**	**WARNING**	**CAUTION**
oral LD_{50}*	up to and including 50 mg/kg	from 50 to 500 mg/kg	greater than 500 mg/kg
inhalation LC_{50}*	up to and including 0.2 mg/liter (0-2,000 ppm)	from 0.2 to 2 mg/liter (2,000-20,000 ppm)	greater than 2 mg/liter (greater than 20,000 ppm)
dermal LD_{50}*	up to and including 200 mg/kg	from 200 to 2,000 mg/kg	greater than 2,000 mg/kg
acute eye effects	corrosive; causes scarring of the cornea and possible clouding of the eye, which cannot be reversed, or irritation that persists for more than 21 days	causes scarring of the cornea and possible clouding of the eye, which can be reversed; irritation may persist for 8–21 days	will not cause scarring of the cornea but can still cause damage; irritation clears in 7 days or less
acute skin effects	corrosive	severe irritation at 72 hours	moderate irritation at 72 hours

Note: *LD_{50} values represent milligrams (mg) of the pesticide per kilogram (kg) of body weight of the test animals. LC_{50} values represent the milligrams of pesticide per liter of air inhaled by the test animals.

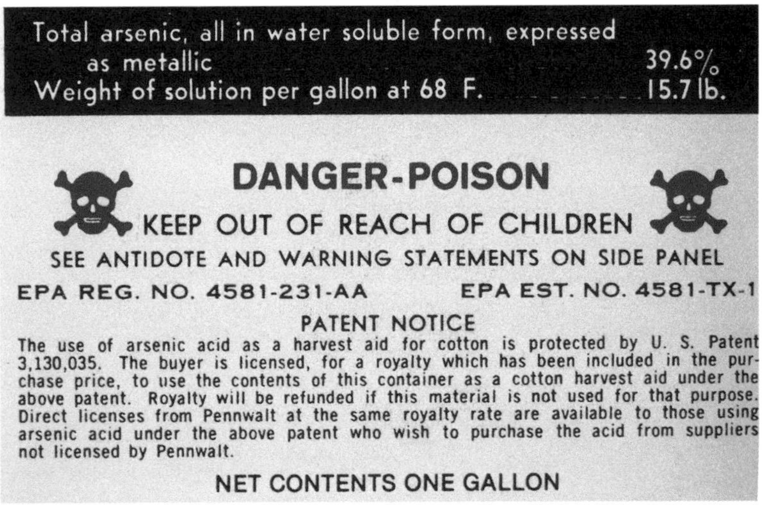

FIGURE 3-1.

The most hazardous pesticides are recognized by the signal words DANGER or DANGER-POISON and a skull and crossbones. A few drops to a teaspoon of these pesticides, taken internally, would probably cause death. These pesticides have an oral LD_{50} of 50 mg/kg or less and a dermal LD_{50} of 200 mg/kg or less.

Signal words are specific to acute human toxicity to the handler. Pesticides labeled DANGER or DANGER-POISON (Fig. 3-1) are the most toxic or hazardous, and regulations normally restrict their use. Category IV pesticides are the least toxic to people and are generally less hazardous. Different label and regulatory requirements apply to each category. For example, you must wear coveralls when mixing and applying DANGER, DANGER-POISON, and WARNING pesticides. Most DANGER pesticides are federally "restricted-use pesticides" that require a certified applicator to use them or supervise their use. Some pesticides are designated "California restricted materials" and are further regulated in California. In the case of these pesticides, you may have to apply for a restricted materials permit or abide by other California regulations to use them legally.

How Pesticides Are Organized

Identify factors that should be considered when selecting pesticides.

Pesticides are organized in a variety of ways to help you easily locate and select an appropriate product for controlling particular pests in particular circumstances. In this section, you will find pesticides sorted by pest target (Table 3-2), chemical family (Table 3-3), mode of action (Table 3-4), and formulation (Table 3-5).

By Pest Target

List groups of pesticides according to pest target and describe the functions of each group.

Pesticides are often organized by the pest they target. Some pesticides can be used to control several different pest types. For instance, petroleum oil products are able to manage weeds, insects, and woody plants like trees or shrubs. Others work only on one pest type. Atrazine, for example, will only control weeds. See Table 3-2 for a list of pesticides (along with brand names) sorted by pest targeted. When you know which pest is damaging your site, having pesticides sorted by pest targeted makes it easier to find the most effective, least damaging pesticide.

TABLE 3-2:

Pesticides organized by pest targeted

Pesticide target	Pesticide type	Example of pesticide chemical and brand name*
algae	algaecide	copper sulfate endothall (Hydrothol 191) sodium hypochlorite (Clorox)
bacteria	bactericide	oxytetracycline (Mycoshield) copper compounds (Basic Copper)
birds	avicide	aminopyridine (Avitrol)
fish	piscicide	rotenone (Prentox)
fungi	fungicide	benomyl (Benlate) copper sulfate (C-O-C-S) chlorothalonil (Bravo)
insects	insecticide	diazinon (Diazinon) permethrin (Ambush) petroleum oils imidacloprid (Gaucho)
mammal predators	predacide	sodium cyanide (M-44)
mites	acaricide	abamectin (Agri-Mek) propargite (Comite)
nematodes	nematicide	1,3-dicloropropene (Telone)
rodents	rodenticide	chlorophacinone strychnine (Gopher Getter) hydroxycoumarin (Warfarin) diphacinone (Diphacin) brodifacoum (Talon) bromadiolone (Maki)
snails and slugs	molluscicide	metaldehyde (Deadline) iron phosphate (Sluggo)
trees and woody shrubs	silvicide	tebuthiuron (Spike) imazapyr (Imazapyr, Arsenal) petroleum oils
weeds	herbicide	simazine (Princep) bromoxynil (Buctril) trifluralin (Treflan) paraquat (Gramoxone) petroleum oil glyphosate (Roundup)

Note: *Some a.i.'s or products listed here may not be currently registered as pesticides or may have had their registration cancelled.

List major chemical families and describe the particular hazards associated with each one.

By Chemical Family

Experts group pesticides according to chemical family. This type of grouping often reveals common characteristics, such as mode of action, chemical structure, and types of formulations possible. There may also be similarities in environmental persistence (how long the pesticide lasts in the environment) and how related pesticides break down through biological processes. Organizing pesticides by chemical family can help you find the best pesticide for a given application. Table 3-3 describes some of the most well-known chemical families and the common chemical names of pesticides with active ingredients that belong to each family.

TABLE 3-3:

Pesticides organized by chemical family

Chemical family	Pesticide types	Chemical characteristics	Mode of action	Common names*
carbamates†	insecticides, fungicides, herbicides, molluscicides, nematicides	some acutely toxic (interfering with human and animal nervous systems); others, like carbaryl, considered to have a low toxicity to people and animals; breaks down rapidly	systemic; enzyme inhibitors	carbaryl, methomyl, metam-sodium, metam-potassium
triazines	herbicides	affects reproductive system of fish, and has negative health effects on humans and animals; chemical accumulates in the liver of mammals; persistent in soil; moves easily with water	systemic; photosynthesis inhibitors	atrazine, simazine, prometryn
pyrethroids†	insecticides	interferes with the function of the nerves and brain in humans and animals; binds tightly to soil and persists in waterway sediment	systemic; interferes with sodium transport in insect nerve cells	bifenthrin, permethrin, cypermethrin
organophosphates†	insecticides, miticides	acutely toxic (interferes with human and animal nervous systems); breaks down rapidly	contact; absorbed through skin, lungs, or digestive tract	malathion, diazinon, chlorpyrifos

Notes: *Some a.i.'s or products listed here may not be currently registered as pesticides, or may have had their registration cancelled.
†These materials are commonly used in animal agriculture in addition to other agricultural situations.

Define mode of action and provide examples of the different modes.

By Mode of Action

People sometimes organize pesticides by their mode of action. A pesticide's mode of action is the way it reacts with a pest organism to destroy it. For instance, an insecticide may act as a growth regulator, an herbicide may prevent photosynthesis, and a fungicide may disrupt spore generation. Table 3-4 lists common pesticides according to their mode of action.

Explain how various modes of action influence pesticide selection.

Understanding mode of action makes it easier to select the right pesticide. It also helps you predict which pesticide works best in a particular situation. For instance, if pests show resistance to one pesticide, a chemical with a different mode of action can reduce the problem (see "Pesticide Resistance" in Chapter 11 for more information).

Usually, pesticides within a chemical class have the same mode of action on specific types of pests. They may also have similar characteristics such as chemical structure, persistence in the environment, and types of formulations possible. Most modes of action fall under one of the following two umbrella terms.

Explain how contact and systemic pesticides control pests differently.

- **Contact.** These pesticides work only on pests that they contact directly. For instance, weeds die when a contact herbicide covers a sufficient surface area of the plant. Only insects that are sprayed directly or have traveled across treated surfaces are affected by contact insecticides.
- **Systemic.** These pesticides work when applied to a particular area of a plant or animal. The pesticide is then translocated, or moved, throughout the organism's system. For example, a systemic herbicide applied to a plant's roots moves throughout the whole plant and kills it. Some insecticides move throughout an insect to kill it after it eats the leaves of a treated plant.

Within these general terms, modes of action become quite specific and can help you effectively select and use a pesticide. Figure 3-2 shows the part of Atrazine 90DF's label that defines its mode of action by its numbered group. Each group number was developed by organizations such as the Insecticide, Herbicide, and Fungicide Resistance Action Committees and the Weed Science Society of America. These organizations are discussed in more detail in Chapter 11.

GROUP 5 HERBICIDE

FIGURE 3-2.

Pesticide labels will have a box like this one, identifying the material's mode of action by group.

TABLE 3-4:

Pesticides organized by mode of action

Group number* and mode of action	Pesticide type	Example of pesticide common and brand name†
	Systemic	
Group 9 – blocks enzyme activity plants need to survive	herbicide	glyphosate (Roundup)
Group 3 – stops plants from growing	herbicide	oryzalin (Surflan)
Group 4 – regulates plant growth	herbicide	2,4-dichlorophenoxyacetic acid (2,4-D)
Group 4a – causes paralysis	insecticide	imidacloprid (Admire)
Group 2 – blocks cell functions (like cell division), causing cell death	fungicide	iprodione (Rovral)
	Contact	
Group 16 – blocks formation of insect exoskeletons	insecticide	buprofezin (Courier)
Group 22 – blocks photosynthesis (photosystem I)	herbicide	paraquat (Gramoxone SL 2.0, Reglone Desiccant)
Group 7 – blocks photosynthesis (photosystem II)	herbicide	propanil (Stam, SuperWham!)
Group M5 – blocks enzyme activity, shuts down cell metabolism	fungicide	chlorothalonil (Bravo)
Group M2 – stops spores from germinating, blocks enzyme activity	fungicide, miticide	sulfur (Microthiol Disperss)

Notes: *Group numbers retrieved from product labels and based on WSSA, IRAC, or FRAC standards.

† Some a.i.'s or products listed here may not be currently registered as pesticides or may have had their registration cancelled.

By Formulation

Explain why pesticides are sold as formulated products.

Pesticide chemicals in their "raw," or unformulated, state are not usually suitable for pest control. These concentrated chemicals (active ingredients) may not mix well with water and may be chemically unstable. For these reasons, manufacturers add other ingredients to improve application effectiveness, safety, handling, and storage. "Other ingredients" are all the substances that manufacturers add to the pesticide active ingredient.

The final product is a pesticide formulation. A formulation consists of

- the pesticide active ingredient
- the carrier, such as an organic solvent or mineral clay
- surface-active ingredients, often including stickers and spreaders
- other ingredients, such as stabilizers, dyes, and chemicals, that improve or enhance pesticidal activity, such as antifreeze to prevent the product from freezing

Usually you need to mix a formulation with water or oil for final application. However, baits, granules, and dusts are ready for use without additional mixing.

The label lists the amount of actual pesticide as percentage of active ingredient (a.i.). Figure 3-3 shows the a.i. of GoalTender as it appears on the label.

Active Ingredient	
oxyfluorfen: 2-chloro-1-(3-ethoxy-4-nitrophenoxy)4-(trifluoromethyl)benzene	41%
Other Ingredients	59%
Total	100%
Contains 4 pounds active ingredient per gallon	

FIGURE 3-3.

Pesticide labels always list the amount of active ingredient in the formulation, as pictured here.

Some labels designate the amount of a.i. in the pesticide name. For example, Diazinon 50W is a dry pesticide containing 50% active ingredient by weight. The 50W in its name tells you that 10 pounds of this formulation will contain 5 pounds of diazinon and 5 pounds of other (also sometimes called inert) ingredients. With liquid formulations, the label lists the pounds of active ingredient in 1 gallon of formulated pesticide. For example, in Lorsban 4E, the "4" indicates that the material contains 4 pounds per gallon of the active ingredient chlorpyrifos.

List the various formulations available and the advantages and disadvantages of each.

Labels usually indicate formulation type by letters that follow or are a part of the brand name of the pesticide. Selecting the right formulation among the many available can be hard. Table 3-5 describes a wide variety of formulation types. See Chapter 11 for methods you can use to select the best formulation for your application.

TABLE 3-5:

Common pesticide formulations

Formulation type	Suffix	Description	Benefits	Drawbacks
wettable powder	W or WP	Wettable powders form a milky suspension in water and consist of the pesticide and a finely ground, dry carrier, usually mineral clay.	Cost and visible residues are reduced when % a.i. is high. Risk of phytotoxicity (plant injury) is reduced. Mixes well with many other pesticides and fertilizers. May be packaged in water-soluble packets to reduce inhalation risk.	Hazards are increased because % a.i. is high. Abrasiveness contributes to pump and nozzle wear. Agitation is always needed to keep it in suspension. Inhalation hazards are increased during mixing and handling because dust particles containing a high % a.i. are fine and can remain suspended in the air for several hours.
dry flowables/ water-dispersible granules	DF WDG	This formulation consists of small granules that must be mixed with water before use.	Less risky to handle because it is less dusty than other formulations. Packaged in easy-to-pour containers, so measuring and mixing is easier than with other formulations.	Abrasiveness contributes to pump and nozzle wear. Agitation is always needed to keep formulation in suspension.

TABLE 3-5:

Common pesticide formulations (continued)

Formulation type	Suffix	Description	Benefits	Drawbacks
soluble powder	S or SP	Similar to wettable powder, except that the pesticide, its carrier, and all its other ingredients completely dissolve in water to form a true solution.	Once dissolved, it needs no additional mixing or agitation. Not abrasive to nozzles or pumps. May be packaged in water-soluble packets to reduce inhalation risk.	Inhalation hazards are increased during mixing and handling because dust particles are fine.
emulsifiable concentrate	E or EC	Emulsifiable concentrates are petroleum-soluble pesticides formulated with emulsifying agents (soaplike materials) and other enhancers. When added to water, they form a milky liquid.	Has many different uses, more than other formulation types. Penetrates porous materials such as soil, fabrics, paper, and wood better than wettable powder. Pours easily for mixing.	Agitation during application is needed to keep the emulsion uniform. Spreads easily when spilled and is hard to clean up. Easily absorbed by porous clothing and leather boots. Passes through the skin more easily than powder formulations. Can cause serious injuries if splashed into eyes. More phytotoxic than other formulation types. Contributes to the breakdown of rubber and plastic parts, some pump parts, and painted surfaces.
flowable	F	A flowable formulation shares many of the characteristics of emulsifiable concentrates and wettable powders. Manufacturers use this formulation when the a.i. will not dissolve in liquids. It combines finely ground pesticide particles with a liquid carrier and emulsifiers.	Easy to handle and apply.	Disadvantages are similar to emulsifiable concentrates. Leaves visible residues. Settles out in the container, so must be shaken thoroughly before mixing.
water-soluble concentrate/ solution	S	This liquid formulation dissolves in water completely.	Does not need agitation after mixing. Not abrasive to application equipment.	Easy to splash and spill when handling.
ultra-low-volume concentrate	ULV	This formulation type has a high a.i. concentration and needs little or no dilution.	Requires less frequent refilling of application equipment. Droplets do not evaporate as rapidly as other formulations' droplets.	Application equipment must be able to apply very small amounts of pesticide over a large area. Equipment calibration must be very accurate.
slurry	SL	A slurry is a thin, watery, pastelike mixture of finely ground dusts. It is usually applied to protect seeds from insects or fungi.	Residues are highly visible, making it easy to see if slurry is evenly distributed.	Powder creates an inhalation hazard when mixing. Needs constant agitation to prevent settling. Abrasiveness contributes to equipment wear.
invert emulsion		A liquid formulation of small water droplets suspended in oil. It dissolves in either oil or water. Invert emulsion concentrates have the consistency of mayonnaise.	Reduces the likelihood of drift. Oil in the formulation reduces runoff and improves rain resistance. Improves surface coverage and absorption because it acts like a sticker-spreader.	Needs continuous agitation. Uses are limited. Regulations prohibit some uses. Hard to get thorough coverage on the undersides of foliage.

Formulation type	Suffix	Description	Benefits	Drawbacks
dust	D	Dusts are made of finely ground pesticide, often combined with a dry carrier that has no chemical action.	Unlikely to damage surfaces where applied. Can provide long-term protection for treated commodities. Used often to control parasites on livestock and poultry and to protect seeds.	Leaves visible residues. Drift is a major hazard. Inhalation hazards are higher than some other formulations. Regulations restrict outdoor applications to periods when the air is still. Application equipment is hard to calibrate. Needs agitation to stop settling and caking in the hopper.
granules	G	Granules are made of a pesticide and carrier combined with a binding agent and are not mixed with water. The most common formulations are in the range of 15 to 30 mesh. *Mesh* is the term used to categorize the size of powder particles based on the number of wires in an inch of screen (the higher the mesh size, the smaller the particles). Granular formulations are more persistent in the environment than other formulations because the pesticide active ingredient releases slowly.	Less likely to drift than other formulation types due to large size. Lessens dust and spray mist hazards to the applicator and environment.	Some need mechanical incorporation into the soil. Often need moisture to activate.
pellets	P or PS	Pellets are identical to granules, except manufacturers mold them into specific uniform weights and shapes. Pellets are applied with equipment such as precision planters to achieve uniformity that is normally hard to accomplish with granules.	Less likely to drift than other formulation types due to size and weight. Uniformity of pellets makes application more precise.	Specialized equipment is needed for proper application.
micro-encapsulated materials		Manufacturers cover liquid or dry pesticide particles in a plastic coating, producing a microencapsulated formulation. Mix microencapsulated pesticides with water and spray them in the same manner as other sprayable formulations. After spraying, the plastic coating breaks down and slowly releases the active ingredient.	Reduces risk for applicators during mixing and application. Release is delayed or slowed, allowing for fewer, less precisely timed applications and longer a.i. effectiveness. Low volatility reduces potential for drift. Less phytotoxic than other formulation types. Less hazardous to the skin than other formulation types.	About the same size as pollen grains, so bees may carry them back to their hives, where capsules break down, releasing the pesticide which poisons the adults and brood. Breakdown sometimes depends on weather conditions, which may result in slower than expected breakdown, leaving higher residues of pesticide in treated areas beyond normal restricted-entry or harvest intervals.
water-soluble bags or packets	WSB or WSP	Manufacturers package preweighed amounts of wettable powder or soluble powder formulations in a special type of plastic bag. As these bags are dropped into the spray tank, they dissolve and release their contents to mix with the water.	Helps reduce mixing and loading hazards of some highly toxic pesticides.	Abrasiveness contributes to equipment wear.

TABLE 3-5:

Common pesticide formulations (continued)

Formulation type	Suffix	Description	Benefits	Drawbacks
baits		Baits are pesticides combined with food, attractants, or feeding stimulants.	Baits attract pests, so pesticides are often put in just a few places.	Baits can be attractive to nontarget organisms and children. Special equipment is sometimes needed to properly apply baits.
impregnates		Used in animal agriculture to control pests. Forms include livestock ear tags, adhesive tapes, medallions, and plastic pest strips.	Ear tags, plastic strips, and medallions allow for pest control over longer periods of time as pesticide transfers from plastic to an animal's hair or coat.	Some impregnates can pose human hazards if they are not handled and applied with care.

Adjuvants

Explain the role of adjuvants in pesticide applications.

Adjuvants are materials you can add to the spray tank to improve pesticide mixing and application or to enhance performance. Manufacturers formulate pesticides to work well under many different application conditions. However, they cannot formulate them for all possible situations. Use adjuvants to customize the formulation to specific needs and local conditions.

Adjuvants are used in order to

- improve the wetting ability of spray solutions
- control evaporation of spray droplets
- improve weatherability of pesticides
- increase the penetration of pesticides through plant or insect cuticles
- adjust the pH of spray solutions
- improve spray droplet deposition
- increase safety to target plants
- correct incompatibility problems
- reduce spray drift

Familiarize yourself with adjuvant types to understand where and how to use them. When choosing adjuvants, outline the effect you wish the adjuvant to have. Next, check pesticide and adjuvant labels to make sure the materials are compatible as well as suited to the application site, target pest, and application equipment. Remember that adjuvants are classified as pesticides, and that their labels are likely to require more personal protective equipment (PPE) than other materials in your tank. Always read and follow all the instructions on an adjuvant's label before mixing.

Often, a single chemical has two or more adjuvant functions. Examples of these are sticker-spreaders and spreader-activators. Some manufacturers also make blends of chemicals for this purpose. These ready-mixed blends are not usually as effective, however, as using several single-active-ingredient adjuvants in amounts customized to your specific needs.

Chapter 3 Review Questions

1. A pesticide is defined as ______________.

- ☐ a. any substance used to control pests in any situation
- ☐ b. only those chemicals registered for pest control in California
- ☐ c. certain pest control products derived from natural sources

2. Match the pesticide group with the pest it controls.

1. acaricide	a. snails
	b. little mallow
2. herbicide	c. ground squirrels
	d. persea mites
3. molluscicide	e. yellow nutsedge
	f. ticks
4. rodenticide	g. slugs
	h. pocket gophers

3. Match the signal word with its oral LD_{50}.

1. DANGER	a. from 50 to 500 mg/kg
2. CAUTION	b. below 50 mg/kg
3. WARNING	c. over 500 mg/kg

4. A pesticide's mode of action is ________________.

- ☐ a. how it will break down once it is released into the environment
- ☐ b. a description of its abrasiveness after mixing
- ☐ c. the method by which it kills or injures the target pest

5. A pesticide formulation is a mixture of ________________.

- ☐ a. concentrated pesticide and adjuvants or other ingredients added to a tank mix
- ☐ b. active ingredient(s) and other ingredients that improve application effectiveness, safety, handling, and storage
- ☐ c. water-soluble packaging and a concentrated chemical that dissolves completely in water

6. Which of the following is *most* important to consider when selecting pesticides for a job?

- ☐ a. the advice of your local pest control adviser, a farm advisor, and the county agricultural commissioner
- ☐ b. target pests, conditions at the application site, and the pesticides' hazards and mode(s) of action
- ☐ c. degree-day calculations, UC IPM's Pest Management Guidelines, and the application equipment available

7. Why are adjuvants used?

- ☐ a. They make mixing and loading safer.
- ☐ b. They prevent groundwater contamination.
- ☐ c. They customize formulations to specific needs.

8. Match the pesticide formulation type with its primary benefit.

1. wettable powder (WP)	a. not abrasive to spray nozzles or pumps
2. dry flowable (DF)	b. reduces the chance of phytotoxicity developing in treated plants
3. soluble powder (SP)	c. less likely to drift than other formulation types
4. emulsifiable concentrates (EC)	d. packaged in easy-to-pour containers that make measuring and mixing easier than other formulations
5. granules	e. has many different uses, more than other formulation types.

CHAPTER 4
LAWS AND REGULATIONS

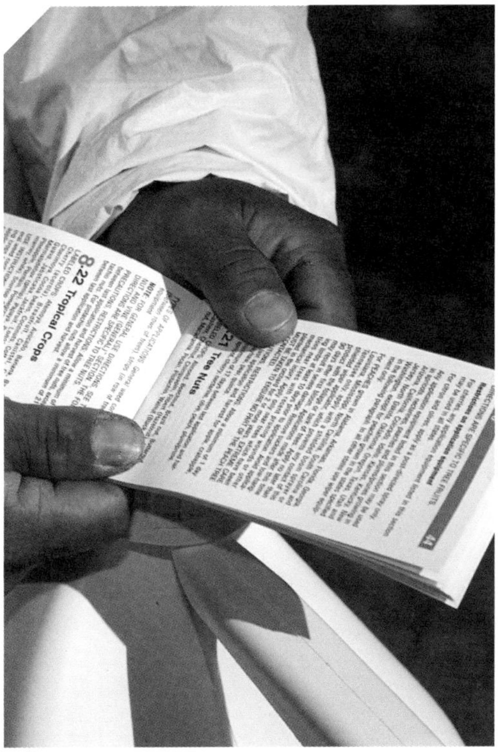

Pesticide Registration and Labeling 42
Chapter 4 Review Questions 53

Knowledge Expectations

1. Explain the legal requirement to read, understand, and follow directions on a pesticide label.
2. Describe the ways you can make labels and Safety Data Sheets readily available to your employees.
3. Identify the information found in the different parts of the label.
4. Describe the type of safety information provided by a pesticide's labeling.
5. Describe the type of safety information provided by a pesticide's Safety Data Sheet.
6. Describe where to find various state and federal laws and regulations that cover applications of restricted-use pesticides on your own land.
7. Describe why, when, and how to obtain a restricted materials permit.
8. List the legal methods you can use to notify farm employees of upcoming pesticide applications and restricted-entry intervals.
9. Describe pesticide use reporting requirements.

Pesticide Registration and Labeling

Explain the legal requirement to read, understand, and follow directions on a pesticide label.

Manufacturers must register pesticides with the U.S. Environmental Protection Agency (EPA) and DPR before they can be used, possessed, or offered for sale in California. These agencies register individual pesticide products. The registration process protects people and the environment from ineffective or harmful chemicals.

Manufacturers must supply labels that meet all federal and state requirements. These labels become legal documents and contain important information for users—the label is the law. It contains all the instructions for how, when, and where you can use the pesticide. Some labels refer to other documents, such as the Worker Protection Standard provisions of Title 40 of the Code of Federal Regulations, part 170 (40 CFR 170). Any documents referred to on pesticide labels become part of the pesticide labeling.

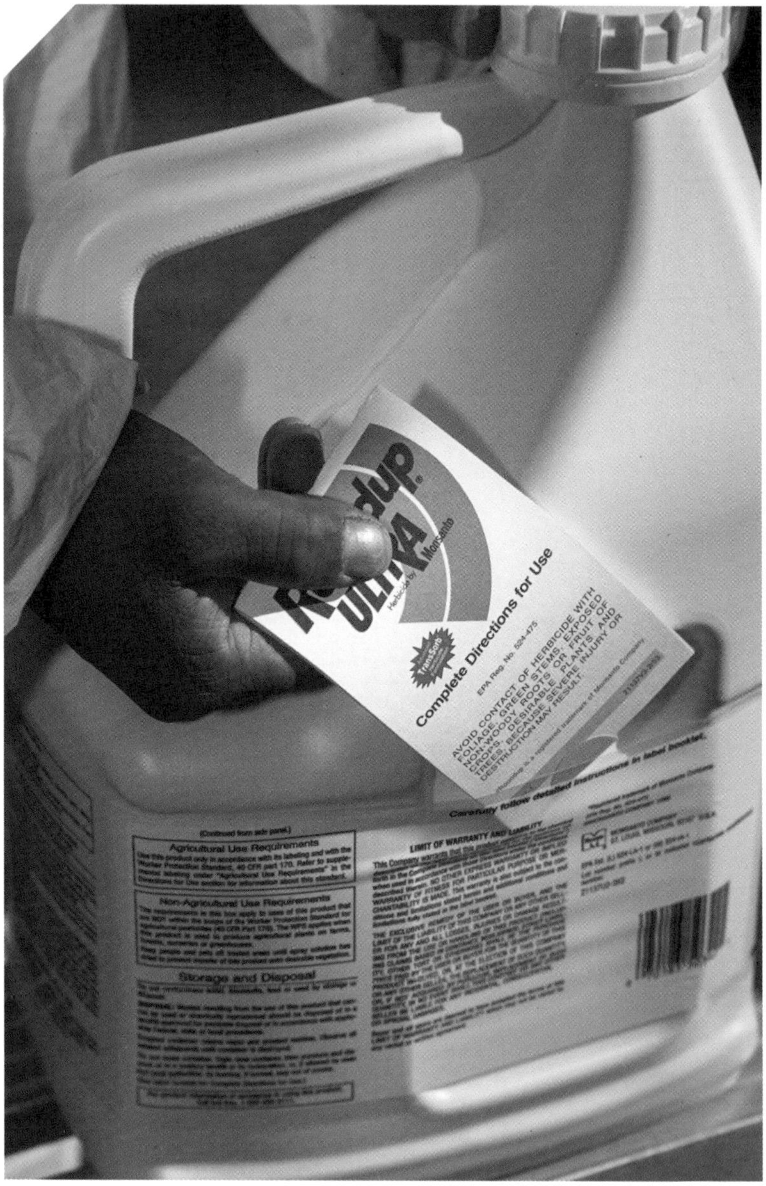

FIGURE 4-1.

Supplemental labels are often attached to pesticide packages. Before purchasing a pesticide, make sure you have a complete set of labels.

Pesticide Labels

The Code of Federal Regulations sets the format for pesticide labels and prescribes the information they must contain. Some packages are too small, however, to have all this information printed on them. In these cases, U.S. EPA requires manufacturers to include the directions for use of the product on accompanying labeling in the form of a booklet, accordion-style foldout, or enclosed leaflet (Fig. 4-1).These booklets, together with the base label, are the complete pesticide label. On metal and plastic containers, manufacturers put these labels in plastic pouches glued to the side of the containers. Paper packages usually have booklets inserted under the bottom flaps. In these cases, the base label attached to the packaging must include a referral statement to the directions for use in the booklet.

California regulations require pesticide labels be available to employees at the use site, so you must ensure labeling is available there.

Describe the ways you can make labels and Safety Data Sheets (SDSs) readily available to your employees.

It is best to keep the complete pesticide label with the pesticide, whether it is in storage, being transported, or being mixed, loaded, and applied, so that employees can readily access the materials. When you must have a copy of a label in two separate locations (say, you return the partially full pesticide container to the storage area but are heading to a different site to apply the mixed pesticide), you can download and print out additional copies of the label from the manufacturer's website.

When to Read the Pesticide Label

Read the pesticide label (Fig. 4-2) carefully

- before buying the pesticide to make sure the pesticide is registered for your intended use; confirm that there are no restrictions or other conditions that prohibit using the pesticide at the application site; check that its use is suitable under current weather conditions; be sure it controls the life stage of your pest; and find out what personal protective equipment (PPE) and special application equipment you must have.
- before mixing and applying the pesticide to learn how to mix and safely apply the material; find out what precautions to take to prevent exposure to people and nontarget organisms; and learn what first aid and medical treatments are necessary should an accident occur.
- when storing pesticides to find out how to properly store the pesticide and learn about any special precautions necessary to prevent fire hazards.
- before disposing of unused pesticide and empty containers to learn how to prevent environmental contamination and hazards to people. (Before disposal, check with the agricultural commissioner in your area for local restrictions and requirements.)

FIGURE 4-2.

The pesticide label is a complex legal document that you must read and understand before making a pesticide application. Make pesticide applications in strict accordance with the label instructions.

What Pesticide Labels Contain

Identify the information found in the different parts of the label.

Describe the type of safety information provided by a pesticide's labeling.

Refer to the corresponding numbers on the sample pesticide label (Fig. 4-4, next page) for examples of the following pesticide label sections.

1. **Statement of Use Classification.** The U.S. EPA classifies pesticides as either general use or restricted use. Federal restricted-use pesticides have a special statement printed on the label in a prominent place (such as the one shown in Figure 4-3). Pesticides that do not contain this statement are general-use pesticides, except where special state restrictions apply. For information, check the DPR list, "California Restricted Materials," which is available from county agricultural commissioners. Some labels have restrictive statements indicating that they are for agricultural or commercial use only. A restrictive statement is different from a statement of use classification.

RESTRICTED USE PESTICIDE
Due to Toxicity to Fish and Aquatic Organisms
For retail sale to and use only by Certified Applicators, or persons under their direct supervision, and only for those uses covered by the Certified Applicator's certification.

FIGURE 4-3.

An example of a restricted-use statement found on the top of the first page of a federal restricted-use pesticide.

2. **Brand Name.** A brand name is the name the manufacturer gives to the product. This is the name used for all advertising and promoting. It is also commonly referred to as the product's trade name.
3. **Ingredients.** Pesticide labels list the percentage of active and other ingredients by weight. Other ingredients are all components of the formulation that do not have pesticidal action. Even if they are not considered active, these ingredients may still be toxic, flammable, or pose other safety or environmental hazards. Some, however, are relatively harmless, such as clay. If a pesticide contains more than one active ingredient, the label will state the percentage of each. Manufacturers do not usually individually identify the names or percentages of other ingredients in the pesticide.

1

RESTRICTED USE PESTICIDE

DUE TO TOXICITY TO NON-TARGET INVERTEBRATES, MAMMALS, AND AQUATIC ORGANISMS

FOR RETAIL SALE TO AND USE ONLY BY CERTIFIED APPLICATORS OR PERSONS UNDER THEIR DIRECT SUPERVISION AND ONLY FOR THOSE USES COVERED BY THE CERTIFIED APPLICATOR'S CERTIFICATION.

2 **Knock 'em out SC**

Miticide/Insecticide By ToxiK™

3 **Active Ingredient:**

4 **Abamectin[1] 8.0%***

Other Ingredients:.......................92.0%

Total: 100.0%

5 [1]CAS No. 71751-41-2

GROUP 6 INSECTICIDE

6 *Knock 'em out SC Miticide/Insecticide is formulated as a suspension concentrate and contains 0.7 lb abamectin per gallon.

7 **EPA Reg. No. 123-456**
EPA Est. 123-CA321

SCP 1351A-L1K 1217

8 **Net Contents**
1 quart (32 fluid ounces)

KEEP OUT OF REACH OF CHILDREN.

9 **WARNING/AVISO**

Si usted no entiende la etiqueta, busque a alguien para que se la explique a usted en detalle. (If you do not understand the label, find someone to explain it to you in detail.)

10 **Manufactured for: ToxiK™, LLC**
P.O. Box 0000 Your Town, Your State, 00000

11

FIRST AID

If swallowed:

- Call a poison control center or doctor IMMEDIATELY for treatment advice.
- Have person sip a glass of water if able to swallow.
- Do not induce vomiting unless told to do so by a poison control center or doctor.
- Do not give anything by mouth to an unconscious person.

If inhaled:

- Move person to fresh air.
- If person is not breathing, call 911 or an ambulance, then give artificial respiration, preferably by mouth-to-mouth, if possible.
- Call a poison control center or doctor for further treatment advice.

If on skin or clothing:

- Take off contaminated clothing.
- Rinse skin IMMEDIATELY with plenty of water for 15-20 minutes.
- Call a poison control center or doctor for treatment advice.

If in eyes:

- Hold eye open and rinse slowly and gently with water for 15-20 minutes.
- Remove contact lenses, if present, after the first 5 minutes, then continue rinsing eye.
- Call a poison control center or doctor for treatment advice.

NOTE TO PHYSICIAN

Early signs of intoxication include dilation of pupils, muscular incoordination, and muscular tremors. Toxicity following accidental ingestion of this product can be minimized by early administration of chemical adsorbents (e.g. activated charcoal).

If toxicity from exposure has progressed to cause severe vomiting, the extent of resultant fluid and electrolyte imbalance should be gauged. Appropriate supportive parenteral fluid replacement therapy should be given, along with other required supportive measures (such as maintenance of blood pressure levels and proper respiratory functionality) as indicated by clinical signs, symptoms, and measurements.

Have the product container or label with you when calling a poison control center or doctor, or going for treatment.

HOT LINE NUMBER

For 24-Hour Medical Emergency Assistance (Human or Animal) or Chemical Emergency Assistance (Spill, Leak, Fire, or Accident), Call **1-800-000-0000**

12

PRECAUTIONARY STATEMENTS

Hazards to Humans and Domestic Animals

WARNING/AVISO

May be fatal if swallowed or inhaled. Do not breathe vapor or spray mist. Harmful if absorbed through the skin. Causes moderate eye irritation. Avoid contact with skin, eyes, or clothing. Remove and wash contaminated clothing before reuse.

Personal Protective Equipment

Applicators and other handlers must wear:

- Long-sleeve shirt and long pants
- Chemical-resistant gloves made of: barrier laminate, butyl rubber ≥ 14 mils, nitrile rubber ≥ 14 mils, neoprene rubber ≥ 14 mils, polyvinyl chloride ≥ 14 mils, or Viton® ≥ 14 mils)
- Shoes plus socks

Discard clothing and other absorbent materials that have been drenched or heavily contaminated with this product's concentrate. DO NOT reuse them. Follow manufacturer's instructions for cleaning/maintaining PPE. If no such instructions for washables exist, use detergent and hot water. Keep and wash PPE separately from other laundry.

Engineering Controls: When handlers use closed systems, enclosed cabs, or aircraft in a manner that meets the requirements listed in the Worker Protection Standard (WPS) for agricultural pesticides [40 CFR 170.240(d)(4-6)], the handler PPE requirements may be reduced or modified as specified in the WPS.

IMPORTANT: When reduced PPE is worn because a closed system is being used, handlers must be provided all PPE specified above for "applicators and other handlers" and have such PPE immediately available for use in an emergency, such as a spill or equipment breakdown.

User Safety Recommendations
Users should:

- Wash hands before eating, drinking, chewing gum, using tobacco, or using the toilet.
- Remove clothing/PPE immediately if pesticide gets inside. Then wash thoroughly and put on clean clothing.
- Remove PPE immediately after handling this product. Wash the outside of gloves before removing. As soon as possible, wash thoroughly and change into clean clothing.

12

PRECAUTIONARY STATEMENTS (CONT.)

Environmental Hazards: This product is toxic to fish and wildlife.

DO NOT apply directly to water, to areas where surface water is present, or to intertidal areas below the mean high water mark. Do not contaminate water when disposing of equipment wash water or rinsate. This product is highly toxic to bees exposed to direct treatment on blooming crops or weeds. DO NOT apply this product or allow it to drift to blooming crops or weeds while bees are foraging in/or adjacent to the treatment area.

Use of this product may pose a risk to threatened and endangered species of fish, amphibians, crustaceans (including freshwater shrimp), and insects. All use of this product in the state of California should comply with the recommendations of the California Endangered Species Project. Before using this product in California, consult with your county agricultural commissioner to determine use limitations that apply in your area.

This product may impact surface water quality due to runoff of rain water. This is especially true for poorly draining soils and soils with shallow groundwater. This product is classified as having a medium potential for reaching both surface water and aquatic sediment via runoff for several weeks to months after application. A level, well-maintained vegetative buffer strip between areas to which this product is applied and surface water features such as ponds, streams, and springs will reduce the potential loading of abamectin from runoff water and sediment. Runoff of this product will be reduced by avoiding applications when rainfall is forecast to occur within 48 hours.

Attention: This product contains a chemical known to the State of California to cause birth defects or other reproductive harm.

13

DIRECTIONS FOR USE

It is a violation of Federal law to use this product in a manner inconsistent with its labeling. Knock 'em out SC must be used only in accordance with recommendations on this label or in separately published TOXIK supplemental labeling recommendations for this product.

14 **DO NOT** apply this product in a way that will contact workers or other persons, or pets either directly or through drift. Only protected handlers are allowed in the area during application. For any requirements specific to your State or Tribe, consult the agency responsible for pesticide regulation.

15

AGRICULTURAL USE REQUIREMENTS

Use this product only in accordance with its labeling and with the Worker Protection Standard (WPS), 40 CFR part 170. This Standard contains requirements for the protection of agricultural workers on farms, forests, nurseries, and greenhouses and handlers of agricultural pesticides. It contains requirements for training, decontamination, notification, and emergency assistance. It also contains specific instructions and exceptions pertaining to the statements on this label about personal protective equipment (PPE) and restricted-entry interval. The requirements in this box only apply to uses of this product that are covered by the Worker Protection Standard.

Do not enter or allow worker entry into treated areas during the restricted-entry interval
16 (REI) of 12 hours. Exception: For grape girdling, cane turning, and tying in grapes, do not enter or allow worker entry into treated areas during the restricted-entry interval (REI) of 4 days.

PPE required for early entry to treated areas that is permitted under the Worker Protection Standard and that involves contact with anything that has been treated, such as plants, soil, or water, is:

- Coveralls over short pants and short-sleeved shirt
- Chemical-resistant gloves made of: barrier laminate, butyl rubber $\geq$ 14 mils, nitrile rubber $\geq$ 14 mils, neoprene rubber $\geq$ 14 mils, polyvinyl chloride $\geq$ 14 mils, or Viton® $\geq$ 14 mils)
- Shoes plus socks

17

STORAGE AND DISPOSAL

Do not contaminate water, food, or feed by storage or disposal.

Pesticide Storage

Store in a tightly closed container in a cool, dry place.

Pesticide Disposal

Pesticide waste may be hazardous. Improper disposal of excess pesticide, spray mixture, or rinsate is a violation of Federal Law. If these wastes cannot be disposed of by use according to label instructions, contact your State Pesticide or Environmental Control agency, or the Hazardous Waste representative at the nearest EPA Regional Office for guidance in proper disposal methods.

Container Handling [less than or equal to 5 gallons]

Non-refillable container. Do not reuse or refill this container. Triple rinse container (or equivalent) promptly after emptying. Triple rinse as follows: Empty the remaining contents into application equipment or a mix tank and drain for 10 seconds after the flow begins to drip. Fill the container 1/4 full with water and recap. Shake for 10 seconds. Pour rinsate into application equipment or a mix tank or store rinsate for later use and disposal. Drain for 10 seconds after the flow begins to drip. Repeat this procedure two more times. Offer for recycling if available or puncture and dispose of in a sanitary landfill, or by incineration, or by other procedures allowed by state and local authorities.

CONTAINER IS NOT SAFE FOR FOOD, FEED, OR DRINKING WATER.

18

CONDITIONS OF SALE AND LIMITATION OF WARRANTY AND LIABILITY

NOTICE: Read the entire Directions for Use and Conditions of Sale and Limitation of Warranty and Liability before buying or using this product. If the terms are not acceptable, return the product at once, unopened, and the purchase price will be refunded.

The Directions for Use of this product must be followed carefully. It is impossible to eliminate all risks inherently associated with the use of this product. Crop injury, ineffectiveness, or other unintended consequences may result because of such factors as manner of use or application, weather or crop conditions, presence of other materials, or other influencing factors in the use of the product, which are beyond the control of TOXIK, LLC or Seller. To the extent permitted by applicable law, Buyer and User agree to hold TOXIK and Seller harmless for any claims relating to such factors.

(*Warranty information continued on supplemental labeling*)

FIGURE 4-4.

This example of a pesticide label shows the most important sections, which are described in the text.

4. **Chemical Name.** Labels must list all chemicals having pesticidal action (active ingredients) in the product. Chemical names describe the active ingredients' chemical structure and are based on international naming rules.
5. **Common Chemical Name or CAS Number.** Chemical names of pesticide active ingredients are often complicated. Therefore, manufacturers give most pesticides common, or generic, names. For example, 0,0-diethyl 0(2-isopropyl-6-methyl-4pyrimidinyl) has the common name diazinon. Not all labels list common names for the pesticide. Some list the CAS number instead, as this label does.
6. **Formulation.** Labels usually list the formulation type, such as emulsifiable concentrate, wettable powder, or soluble powder. Manufacturers may include this information as a suffix in the brand name of the pesticide. For example, in the name Princep 80W, the "W" indicates a wettable powder formulation. (See Table 3-5 in Chapter 3 for definitions of many suffixes used with brand names.)
7. **Registration and Establishment Numbers.** The U.S. EPA assigns registration numbers to each pesticide. In addition, an establishment number identifies the site of manufacture or repackaging. If the product requires registration in California (but not with U.S. EPA), the Department of Pesticide Regulation will assign a California registration number.
8. **Contents.** Labels list the net contents, by weight or liquid volume, contained in the package.
9. **Signal Word.** An important part of every label is the signal word. The words DANGER and POISON (with a skull and crossbones) indicate that the pesticide is highly toxic. The word DANGER used alone indicates that the pesticide poses a dangerous health hazard. WARNING indicates moderate toxicity, and CAUTION means low toxicity (see "Pesticide Toxicity Categories" in Chapter 3). During the registration process each pesticide is assigned a toxicity category (Category I, DANGER, to Category IV, no signal word required). The level of hazard determines the signal word manufacturers must use on their labels.
10. **Manufacturer.** Pesticide labels always contain the name and address of the manufacturer of the product. Use this address if you need to contact the manufacturer for any reason.
11. **First Aid.** The first aid statement provides emergency information. It tells what to do to decontaminate someone who becomes exposed to the pesticide. It describes the emergency first aid procedures for swallowing, skin and eye exposure, and inhalation of dust or vapors. This section tells you when to seek medical attention.
12. **Precautionary Statements.** Precautionary statements describe the pesticide hazards. Read and follow the instructions given in a precautionary statement. The statement includes as many as three areas of hazard. The most important hazards are those to people and domestic animals.

    The first part of a precautionary statement explains why the pesticide is hazardous, lists adverse effects that may occur if people become exposed, and describes the type of PPE to wear while handling containers and while mixing and applying the product.

    The second part of a precautionary statement describes environmental hazards. It tells you whether the pesticide is toxic to nontarget organisms such as honey bees, fish, birds, and other wildlife. Here is where you learn how to avoid environmental contamination.

    The third part of the precautionary statement explains special physical and chemical hazards. These include risks of fire or explosion and hazards from fumes.
13. **Directions for Use.** The directions for use are an important part of the pesticide label. It is a violation of the law if you do not follow these instructions. The only exceptions are cases where federal or state laws specify acceptable deviations from label instructions. The directions for use list all the target pests that manufacturers claim their pesticide controls. It also includes the crops, plant species, animals, or other sites to which you can apply the pesticide. Here is where you find special restrictions that you must observe. These include crops that you may or may not plant in the treated area (plantback restrictions, also called

rotational crop restrictions). They also include restrictions on feeding crop residues to livestock or grazing livestock on treated plants. These instructions also tell you how to apply the pesticide (including allowable application methods) and provide methods to help you prevent drift. They specify how much pesticide to use, where to use the material, and when to apply it (Fig. 4-5). The directions include the harvest intervals (or preharvest intervals) for all crops whenever appropriate. A harvest interval is the time, in days, required after application before you may harvest an agricultural crop.

Explain the legal requirement to read, understand, and follow directions on a pesticide label.

BRASSICA LEAFY VEGETABLES CROPS AND TURNIP GREENS

All members of the Brassica Leafy Vegetable Group 5, plus Turnip greens, including: Broccoli, Broccoli raab (rapini), Brussels sprouts, Cabbage, Cauliflower, Cavalo broccolo, Chinese broccoli (gai lon), Chinese cabbage (bok choy), Chinese cabbage (napa), Chinese mustard cabbage (gai choy), Collards, Kale, Kohlrabi, Mizuna, Mustard greens, Mustard spinach, Rape greens, Turnip greens

PEST		QUARTS OF THIS PRODUCT PER ACRE	SPECIFIC DIRECTIONS
Flea beetles Harlequin bug Leafhoppers		1/2 to 1	Repeat applications as needed up to a total of 4 times per year but not more often than once every 7 days.
Armyworm Aster leafhopper Corn earworm Diamondback moth Fall armyworm Imported cabbageworm	Lygus bugs Spittle bugs Stink bugs Tarnished plant bug	1 to 2	

FIGURE 4-5.

Many pesticide labels have tables like this one, which shows application rates and directions to control listed pests on specified crops. These tables are found in the "Directions for Use" section of the pesticide label.

14. **Misuse Statement.** The misuse statement reminds users to apply pesticides according to label directions.
15. **Agricultural Use Requirements.** This special statement appears in the "Directions for Use" section on labels of pesticides approved for use in production agriculture, commercial greenhouses and nurseries, and forests. It refers to the Worker Protection Standard (40 CFR 170). You must use the pesticide according to this standard as well as the requirements on the pesticide label. It provides information on the PPE required for early-entry workers. It also gives the restricted-entry interval for workers (see no. 16, below).
16. **Restricted-Entry Statement.** Usually a period of time must elapse before anyone can enter a treated area unless they are wearing PPE. This period is the restricted-entry interval. Restricted-entry intervals may vary according to the toxicity and special hazards associated with the pesticide. The crop or site being treated, and its geographic location, also influence the length of this interval. Some pesticide uses in California require longer restricted-entry intervals than those listed on the pesticide label. Check with the local county agricultural commissioner for this information.
17. **Storage and Disposal Directions.** This section contains directions for properly storing and disposing of the pesticide and empty pesticide containers. Proper disposal of unused pesticides and pesticide containers reduces human and environmental hazards. Some pesticides have special storage requirements because improper storage causes them to lose their effectiveness. Improper storage may even cause explosions or fires.
18. **Warranty.** Manufacturers usually include a warranty and disclaimer on their pesticide labels. This information informs you of your rights as a purchaser and limits the liability of the manufacturer.

Describe the type of safety information provided by a pesticide's Safety Data Sheet.

Safety Data Sheets

In addition to the label, you will also want to review a pesticide's Safety Data Sheet (SDS), formerly known as a Material Safety Data Sheet, or MSDS. The SDS provides detailed information about pesticide hazards (Fig. 4-6). Information found on an SDS includes (but is not limited to)

- the chemical characteristics of active and other hazardous ingredients
- fire and explosion hazards
- health hazards
- reactivity and incompatibility characteristics
- storage information
- emergency spill or leak cleanup procedures
- LD_{50} and LC_{50} ratings for various test animals
- emergency telephone numbers of the manufacturer

Manufacturers prepare these sheets and make them available to every person selling, storing, or handling pesticides. Ask your employer for them, or, if self-employed, obtain them from the chemical manufacturer or pesticide supplier. You can obtain SDSs for every labeled pesticide, and employers are required to keep them in a clearly labeled, accessible area.

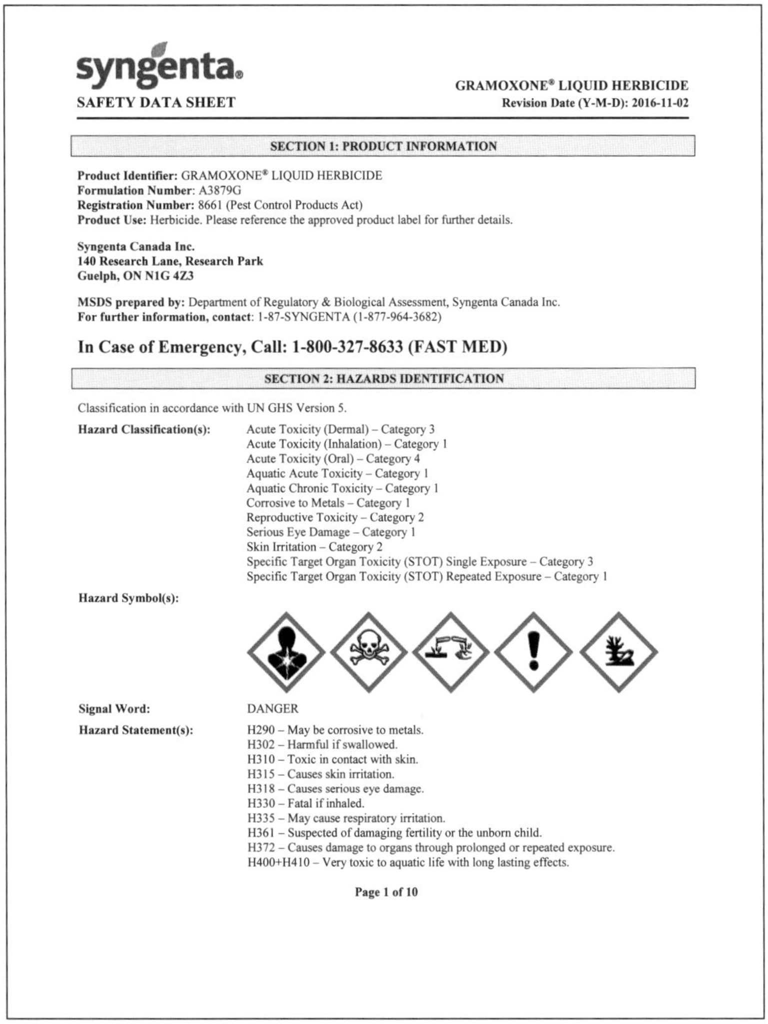

syngenta.
SAFETY DATA SHEET

GRAMOXONE® LIQUID HERBICIDE
Revision Date (Y-M-D): 2016-11-02

SECTION 1: PRODUCT INFORMATION

Product Identifier: GRAMOXONE® LIQUID HERBICIDE
Formulation Number: A3879G
Registration Number: 8661 (Pest Control Products Act)
Product Use: Herbicide. Please reference the approved product label for further details.

Syngenta Canada Inc.
140 Research Lane, Research Park
Guelph, ON N1G 4Z3

MSDS prepared by: Department of Regulatory & Biological Assessment, Syngenta Canada Inc.
For further information, contact: 1-87-SYNGENTA (1-877-964-3682)

In Case of Emergency, Call: 1-800-327-8633 (FAST MED)

SECTION 2: HAZARDS IDENTIFICATION

Classification in accordance with UN GHS Version 5.

Hazard Classification(s):
Acute Toxicity (Dermal) – Category 3
Acute Toxicity (Inhalation) – Category 1
Acute Toxicity (Oral) – Category 4
Aquatic Acute Toxicity – Category 1
Aquatic Chronic Toxicity – Category 1
Corrosive to Metals – Category 1
Reproductive Toxicity – Category 2
Serious Eye Damage – Category 1
Skin Irritation – Category 2
Specific Target Organ Toxicity (STOT) Single Exposure – Category 3
Specific Target Organ Toxicity (STOT) Repeated Exposure – Category 1

Hazard Symbol(s):

Signal Word: DANGER

Hazard Statement(s):
H290 – May be corrosive to metals.
H302 – Harmful if swallowed.
H310 – Toxic in contact with skin.
H315 – Causes skin irritation.
H318 – Causes serious eye damage.
H330 – Fatal if inhaled.
H335 – May cause respiratory irritation.
H361 – Suspected of damaging fertility or the unborn child.
H372 – Causes damage to organs through prolonged or repeated exposure.
H400+H410 – Very toxic to aquatic life with long lasting effects.

Page 1 of 10

FIGURE 4-6.

This excerpt of a pesticide's Safety Data Sheet illustrates several sections that can sometimes have different information than is on the product's label (e.g., the signal word on an SDS may be different from the label's signal word). The sections pictured include product identification, hazards identification, and regulatory information. Always use the pesticide in accordance with the information provided on the pesticide label.

Sections of a Safety Data Sheet

Changes to Occupational Safety and Health Administration (OSHA) regulations have standardized the topics and format of SDSs. These standards mandate that all SDSs contain the following 16 sections.

1. Identification. Identification of the product includes the brand name of the product as it appears on the label, as well as any alternate ways to identify the product, such as other trade names or synonyms, chemical name(s), or the U.S. EPA registration number. This section also outlines the product's recommended uses and restrictions on use and provides you with the name, address, and telephone number of the manufacturer, along with (on some labels) a mode of action classification. Emergency phone numbers are also printed here.

2. **Hazards Identification.** This section must state the signal words, along with all relevant symbols or symbol descriptions (Fig. 4-7) and precautionary statements. This section must also include additional descriptions of hazards that have been identified during the classification process but do not inform the actual class of the pesticide. Information about ingredients that have not been tested for acute toxicity may also be included here in certain circumstances.
3. **Composition/Information on Ingredients.** Information about the composition of the pesticide is included in this section. Each classified ingredient, additive, and impurity is listed, along with common names and synonyms, CAS (Chemical Abstracts Service) numbers, or other unique identifiers, and the percentage of each ingredient in the undiluted formulation. You may also see a note about any ingredients that are not revealed because of a trade secret claim.
4. **First Aid Measures.** Statements in this section provide information needed to assess and respond to exposure incidents, including the various ways people get exposed, major symptoms (both acute and delayed), and the immediate steps you can take to treat the exposed person. Special indications for medical treatment are also found in this section.
5. **Firefighting Measures.** This section describes the pesticide's potential to create fire hazards, including hazardous chemicals that may be released during a fire. It includes a list of suitable extinguishing media (e.g., water, foam, dry powder) and, where necessary, unsuitable extinguishing media. In addition, you will find information specifically for firefighters, such as a list of special protective equipment and instructions for avoiding environmental contamination.

FIGURE 4-7.

Hazard classes and pictograms that may be used in section 2 of a Safety Data Sheet, "Hazards Identification." Pictograms communicate specific hazards associated with the pesticide.

GHS Pictograms and Hazard Classes		
 Oxidizers	 Flammables Self-Reactives Pyrophonics Self-Heating Emits Flammable Gas Organic Peroxides	 Explosives Self-Reactives Organic Peroxides
 Acute Toxicity (Severe)	 Corrosives	 Gases under Pressure
 Carcinogen Respiratory Sensitizer Reproductive Toxicity Target Organ Toxicity Mutagenicity Aspiration Toxicity	 Environmental Toxicity	 Irritant Dermal Sensitizer Acute Toxicity (Harmful) Narcotic Effects Respiratory Tract Irritation

6. **Accidental Release Measures.** During a pesticide spill (accidental release), check this section to find out what PPE, precautions, and procedures you should use to keep the spill contained and begin cleanup. This section lists recommended materials for cleaning up the pesticide and details disposal options.
7. **Handling and Storage.** An explanation of the precautions you must take to handle the pesticide safely can be found here. In addition, you can find a description of safe storage conditions for the pesticide.
8. **Exposure Controls/Personal Protection.** This section provides the OSHA permissible exposure limit (PEL) and any other exposure limit used by the manufacturer to describe the maximum allowable exposure. It also lists appropriate engineering controls, personal protection measures, and PPE required to keep you from exposure to the active ingredient or other ingredients that pose health risks.
9. **Physical and Chemical Properties.** Where they are known, 17 physical and chemical properties of the pesticide are listed in this section. Properties described include the appearance of the formulation, its odor, flammability or explosive limits, solubility, and viscosity, among others.
10. **Stability and Reactivity.** The pesticide's stability and reactivity information is listed in this section. Here you will find a list of chemicals or conditions that create instability, reactivity, or incompatibility during mixing and storage. You can find out how likely it is that you will experience hazardous reactions, as well as the conditions to avoid (like static discharge, exposure to extreme heat, shock, or vibrations) to reduce bad reactions. It also lists hazardous decomposition products to help you avoid unintended exposure to harmful substances.
11. **Toxicological Information.** This section details the chronic and acute effects of pesticides through the four likely routes of exposure (inhalation, ingestion, skin, and eye contact) for both long- and short-term exposures. Here you will find numerical measures of toxicity (such as LD_{50} estimates based on animal studies), as well as symptoms related to the pesticide's physical, chemical, and toxicological characteristics. If the pesticide has been found to be a potential carcinogen, that information will be listed here as well.
12. **Ecological Information (Nonmandatory).** In this section, you may see the pesticide's toxicity to sensitive areas and organisms. For instance, the LD_{50} for honey bees may be listed, as well as information about the pesticide's half-life in soil or in water at certain temperatures. You may also find information about the pesticide's mobility in soil, its potential to bioaccumulate, and other adverse effects it can have on the environment.
13. **Disposal Considerations (Nonmandatory).** Detailed instructions can be found in this section for safe handling and proper disposal of waste residues, pesticide containers, and contaminated packaging.
14. **Transport Information (Nonmandatory).** This section may include any information that relates to the local or long-distance transport of the pesticide. For instance, you may see Department of Transportation or other organization's transport hazard class(es) named here, as well as a listing of the pesticide's major environmental hazards. You may also find special precautions for users who will be transporting the pesticide from one place to another, whether moving it from one location to another on the same property or from one town to another.
15. **Regulatory Information (Nonmandatory).** In this section, you will see safety, health, and environmental regulations that pertain to the pesticide. For pesticides registered for use in California, you may see a Proposition 65 warning message.
16. **Other Information.** This section contains information the manufacturer deems important but which does not fit into any of the other sections. It can include revision history or the date of SDS preparation and can also include a brief discussion of the manufacturer's liability or other information about the product (see OSHA 2012).

Regulations Covering the Handling of Restricted Materials on Your Own Land

Describe where to find various state and federal laws and regulations that cover applications of restricted-use pesticides on your own land.

State and federal regulations exist that dictate how you may use restricted materials on land you own or manage. These regulations include the requirement that you take specific certification exams to prove that you understand how to handle pesticides safely. In addition, you must obtain a permit from your local county agricultural commissioner to apply California restricted materials, even if you are applying them on your own property. Your permit may contain additional local requirements for the use of specific restricted materials. Regulations covering pesticide use in California can be found in Title 3, Division 6, of the California Code of Regulations, and they are considered equivalent to federal regulations. See the Department of Pesticide Regulation website, cdpr.ca.gov/docs/legbills/calcode/chapter_.htm, to review these regulations.

Obtaining a Restricted Materials Permit

Describe why, when, and how to obtain a restricted materials permit.

Before applying any state or county restricted material on your property, you are required to obtain a restricted materials permit. To get a permit, a property owner or business operator must apply to their local county agricultural commissioner. The application must list the areas to be treated, their location and size, crops or commodities, pest problems, names of California restricted materials that are being requested to be applied, and application method, and it must confirm that alternative mitigation methods have been considered. If you own or manage several locations in a single county (for example, different fields), they can all be covered with a single permit so long as you clearly identify and describe each site.

Your permit application must also include a map or description of the surrounding area showing any places that could be harmed by pesticides. These could include rivers, schools, hospitals, labor camps, residential areas, endangered species habitats, and nearby susceptible livestock or crops.

Notification Requirements

List the legal methods you can use to notify farm employees of upcoming pesticide applications and restricted-entry intervals.

You must notify employees working on your farm of any upcoming pesticide applications and restricted-entry intervals that will be in effect if they will be working within ¼ mile of the treated area(s). You need to tell them the treatment location, timing, pesticide name, and any precautions they should take. Pesticide labels may specify the method of notification you must use, such as posting or oral notification, or both. If no method is specified, you may notify workers either orally or by posting. California regulations require field posting if the restricted-entry interval exceeds 48 hours. For an enclosed space (such as a hoophouse), regulations require the posting of notification if the restricted-entry interval exceeds 4 hours, except if the space is entirely enclosed (such as a greenhouse). In that case, the area must be posted regardless of the length of the restricted-entry interval (Fig. 4-8). In every case, you or your employer are required to *display* application-specific information as well as the pesticide's SDS at a central location easily available to employees who will be entering treated fields.

FIGURE 4-8.

Labels may require the posting of warning signs like this one to stop people from entering a recently treated area. Signs must contain a skull and crossbones at the center and the word DANGER in letters big enough to be read from a distance of 25 feet.

Application-specific information and the related SDS must be displayed for a total of 30 days plus the length of the restricted-entry interval after a pesticide has been applied to a site. After that time, records of pesticide applications by site and the labels and SDSs of these pesticides must be stored in a place where they can be made available to employees upon request. See Pesticide Safety Information Series (PSIS) A-8 and A-9 for information about posting, and be sure to display these leaflets at a central location and at permanent decontamination sites servicing 11 or more fieldworkers or handlers. Both PSIS A-8 and A-9 are available in various languages at the Department of Pesticide Regulation's website, cdpr.ca.gov.

Use Reporting Requirements

Describe pesticide use reporting requirements.

You must submit a pesticide use report (PUR) to your county agricultural commissioner each month for all agricultural pesticides you have applied. The PUR must be submitted by the 10th day of the month following the pesticide application. When you get your permit, your agricultural commissioner will explain what needs to be reported after pesticide applications are made on your property and how to deliver this information. The county agricultural commissioner then reports that data to the Department of Pesticide Regulation. A sample PUR form can be found in Sidebar 4-1.

SIDEBAR 4-1

Sample Pesticide Use Report (PUR) Form

This PUR form provides an example of what you will fill out to make your monthly pesticide use reports. Since these forms are often customized by local agricultural commissioners' offices, yours may look slightly different from the one pictured here.

STATE OF CALIFORNIA
MONTHLY SUMMARY PESTICIDE USE REPORT
DPR-PML-060 (REV. 1/18) PAGE 1 OF 2

DEPARTMENT OF PESTICIDE REGULATION
PEST MANAGEMENT AND LICENSING BRANCH

INSTRUCTIONS FOR COMPLETING THIS FORM ARE INDICATED BELOW AND ON THE REVERSE SIDE

OPERATOR (FIRM NAME)	ADDRESS	CITY	ZIP CODE	PHONE NUMBER

OPERATOR ID/PERMIT NUMBER	LICENSE NUMBER	COUNTY WHERE APPLIED	COUNTY NUMBER	MONTH/YEAR OF USE	TOTAL NUMBER OF APPLICATIONS

1. Complete Columns A, B, C, and D for All Users
2. Complete Column E by using one of the following codes:
Code 10 - Structural Pest Control.....*includes any pest control work performed within or on buildings and other structures.*
Code 30 - Landscape Maintenance Pest Control.....*includes any pest control work performed on landscape plantings around residences or other buildings, golf courses, parks, cemeteries, etc.*
Code 40 - Right-of-Way Pest Control.....*includes any pest control work performed along roadsides, power lines, median strips, ditch banks, and similar sites.*
Code 50 - Public Health Pest Control.....*includes any pest control work performed by or under contract with State or local public health or vector control agencies.*
Code 80 - Vertebrate Pest Control.....*includes any vertebrate pest control work performed by public agencies or work under the supervision of the State or county agricultural commissioner.*
Code 91 - Commodity Fumigation (Nonfood/Nonfeed).....*includes fumigation of nonfood/nonfeed commodities such as pallets, dunnage, furniture, burlap bags, etc.*
Code 100 - Regulatory Pest Control.....*includes any pest control work performed by public employees or contractors in the control of regulated pests.*
Note: pesticide use on cannabis should be reported on the appropriate production agricultural pesticide use report forms (DPR-PML-017C, DPR-PML-183 or DPR-PML-025).
3. Complete Columns F and G, if use does not fit one of the above codes

A	B	C	D	E	F	G
MANUFACTURER AND NAME OF PRODUCT APPLIED	EPA/CALIFORNIA REGISTRATION NUMBER FROM LABEL INCLUDE ALPHA CODE	TOTAL PRODUCT USED (Check One Unit of Measure)	NUMBER OF APPLICATIONS	CODE	COMMODITY OR SITE TREATED	ACRES/UNITS TREATED
		O LB O OZ O PT O QT O GA				
		O LB O OZ O PT O QT O GA				
		O LB O OZ O PT O QT O GA				
		O LB O OZ O PT O QT O GA				
		O LB O OZ O PT O QT O GA				
		O LB O OZ O PT O QT O GA				

REPORT PREPARED BY ______________________ DATE ______________

Distribution: CAC - Two copies; Report preparer - One copy

Chapter 4 Review Questions

1. California regulations require field posting if the restricted-entry interval exceeds _____.

☐ a. 48 hours
☐ b. 24 hours
☐ c. 12 hours

2. What documents are considered part of a pesticide label?

☐ a. any documents that are posted in your workplace
☐ b. any documents that are referred to on the label
☐ c. any documents that are provided by pesticide dealers

3. Match the section of the label with the information found there.

1. Directions for Use	a. when to seek medical attention after exposure to the product
2. First Aid	b whether or not the product is toxic to honey bees
3. Precautionary Statement	c. the list of all target pests that the manufacturer claims the product controls
	d. the amount of time that must go by before anyone can enter a treated area unless they are wearing PPE
4. Restricted-Entry Statement	e. plantback or rotational crop restrictions
	f. what PPE to wear when handling containers and while mixing and applying the product

4. Match the information needed with the best source for it.

1. how to clean up after a pesticide spill	a. label
2. how to apply the pesticide correctly	
3. how to dispose of pesticide containers	b. Safety Data Sheet
4. how to avoid hazardous reactions	

5. True or false?

☐ True ☐ False a. Restricted materials permits are issued by the Department of Pesticide Regulation's Pest Management and Licensing Branch.
☐ True ☐ False b. The label is the law.
☐ True ☐ False c. Employees must have access to pesticide use records for treated fields and for materials they handle.
☐ True ☐ False d. You must report all agricultural pesticide use to your county agricultural commissioner once a year.
☐ True ☐ False e. You have to have a restricted materials permit before you are allowed to apply any California restricted material on your property.
☐ True ☐ False f. Regulations covering pesticide use in California can be found in Title 3, Division 6, of the California Code of Regulations.

CHAPTER 5
ENVIRONMENTAL HAZARDS

The Environment and the Hazards of Pesticide Use............ 56
Pesticide Characteristics 56
How Pesticides Behave in the Environment 57
Environmental Impacts of Pesticide Application................. 61
Chapter 5 Review Questions 64

Knowledge Expectations

1. Explain the potential environmental hazards associated with pesticides.
2. Describe pesticide characteristics and how they influence the potential for pesticides to move offsite.
3. List the types of offsite movement of pesticides.
4. Distinguish between point sources and nonpoint sources of environmental contamination by pesticides.
5. Describe factors that influence offsite movement of pesticides.
6. List features of a given site that influence the potential for a pesticide to reach surface water or groundwater.
7. Explain how pesticide residues can build up on agricultural commodities.
8. Identify toxicity and residue potential of pesticides that are commonly applied to animals or animal product agricultural commodities.
9. Describe ways that pesticides can impact nontarget organisms.

The Environment and the Hazards of Pesticide Use

Explain the potential environmental hazards associated with pesticides.

California has been active in pesticide regulation since passing its first pesticide law in 1901. California's Department of Pesticide Regulation (DPR) and county agricultural commissioners work with the U.S. Environmental Protection Agency (EPA) to regulate pesticide use. These agencies face an increasing challenge: protect the public, workers, and the environment while allowing growers to manage pests using pesticides. In California, regulators make sure we have safe and sensible pesticide rules, and they also make sure pesticide users follow those rules.

The environmental hazards covered in this chapter include the contamination of surface water (lakes, streams, irrigation ditches, etc.) and groundwater (aquifers), the damage to nontarget organisms (e.g., pollinators, endangered species, wildlife in natural habitats, etc.), and the contamination of sensitive areas (e.g., schools, apiaries, domestic animal habitat, wildlands, homes, etc.). Related to these topics are the various ways pesticides escape from treated areas and enter the environment, such as drift, leaching, runoff, and via residues that move into the environment before the pesticide breaks down. In addition, the factors that influence offsite movement of pesticides will be covered.

The environment is made up of everything around us. It includes not only the natural elements that the word *environment* most often brings to mind, but also people and the manufactured components of our world. Anyone who uses a pesticide must consider how that pesticide affects the environment.

Pesticide Characteristics

Describe pesticide characteristics and how they influence the potential for pesticides to move offsite.

To understand how pesticides move in the environment, you must first understand the different physical and chemical characteristics of pesticides and how they change a pesticide's interaction with the environment. These characteristics are

- solubility: the ability of a pesticide to dissolve in a liquid; soluble pesticides are more likely to move with water in surface runoff or move in water through the soil (leaching) than are less water-soluble pesticides, like those that dissolve in oil (Fig. 5-1)
- adsorption: the process whereby a pesticide binds to soil particles; a pesticide that adsorbs to soil particles is less likely to leach than a chemical that does not adsorb tightly to the soil, but it can still move offsite via soil erosion
- persistence: the length of time that a pesticide will remain present and active in its original form before breaking down; the longer it takes for a pesticide to break down, the longer the pesticide remains in the environment (Fig. 5-2)
- volatility: the tendency of a pesticide to turn into a gas or

Vapor Pressure: Paraquat Dichloride 7.5 x 10(-8) mmHg @ 77°F (25°C)
Vapor Density: Not available
Relative Density: 1.07 - 1.13 g/ml @ 68°F; 9.12 lbs/gal
Solubility (ies): Paraquat Dichloride 620 g/l @ 68°F (20°C)

FIGURE 5-1.

Section 9 of a pesticide's SDS lists the material's chemical properties. Here you can see the conditions under which this pesticide's active ingredient will dissolve in water.

Persistance and degradability

Oxyfluorfen

Biodegradability: Material is expected to biodegrade very slowly (in the environment). Fails to pass OECD/EEC tests for ready biodegradability.

Theoretical Oxygen Demand: 1.305 mg/mg

Stability in Water (1/2-life)
Hydrolysis, 3.9 d, pH 5 - 9, Half-life Temperature 20°C

FIGURE 5-2.

An example of an SDS statement describing the persistence of the pesticide in the environment.

9. PHYSICAL AND CHEMICAL PROPERTIES

Appearance: Grey-tan granule
Loose bulk Density: 41 - 56 lb/cu ft.

Solubility in H_2O
Active Ingredient – 375 mg/l (25°C) Miscible with common organic solvents
Vapor Pressure
Active Ingredient – 10 x 10^3 mPa (25°C)

FIGURE 5-3.

An example of an SDS statement describing the volatility of a pesticide. Highly volatile pesticides have vapor pressure values above 1 x 10^{-4}. This pesticide has a vapor pressure rating of 10 x 10^3. It is so volatile that you are required to incorporate it into the soil directly after application.

vapor; the chance of volatilization increases as temperatures and wind increase and when relative humidity is low (Fig. 5-3)

How Pesticides Behave in the Environment

List the types of offsite movement of pesticides.

Environmental contamination from pesticides can occur in a number of ways (Fig. 5-4). It may be the result of drift, when wind and air currents carry pesticides away from the application site. It can also result when pesticides you have applied run off over land into surface water sources or when persistent pesticides leach into groundwater. Pesticides can move away from the application site on or in objects, plants, or animals as well.

Sometimes, environmental damage can occur even when the pesticides you have applied remain in the target area. For example, if nontarget species are in a field under treatment or enter such a field soon after an application, they may be poisoned. Some pesticides are so persistent

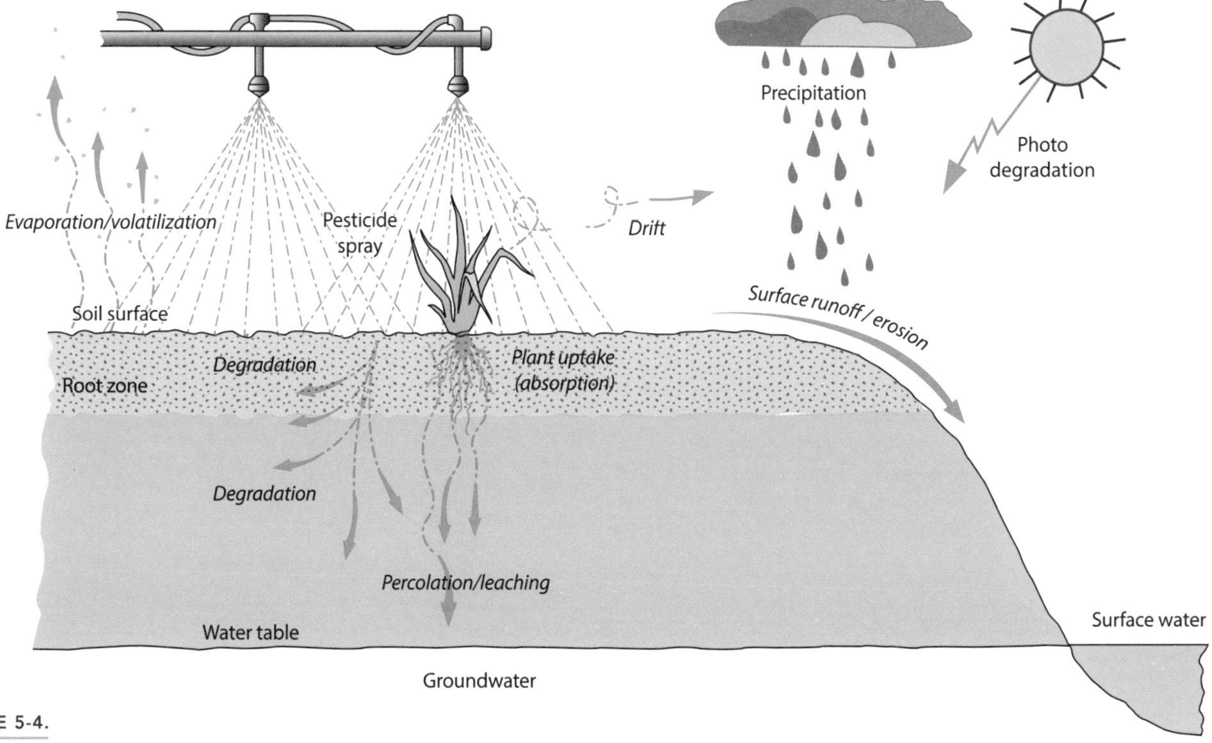

FIGURE 5-4.

The various fates of pesticides in the environment. *Source:* Randall 2008.

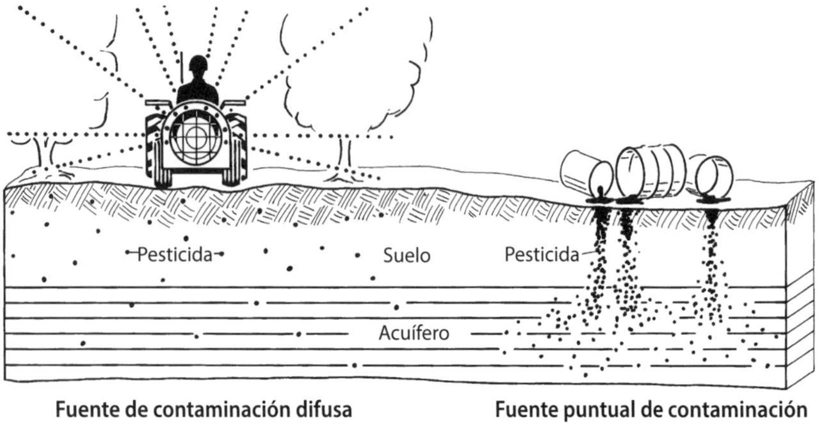

FIGURE 5-5.

Point pollution sources are areas where large quantities of pesticide or other pollutants are discharged into one location. Nonpoint pollution sources arise from normal application of pesticide or other material over a large area.

that they remain in the environment for many years after they have been applied. Therefore, areas that were previously used for growing crops and have since been converted into other uses may still have residues of these pesticides.

Environmental contamination can also occur through point source and nonpoint source pollution (Fig. 5-5). Spilling or dumping pesticides or rinse water repeatedly in a specific place, such as near equipment cleanup sites, in or around pesticide storage areas, or in places where pesticides are regularly mixed and loaded, is called point source pollution. Nonpoint source pollution comes from applications made over a widespread area. The movement of pesticides into streams or groundwater in rain or irrigation water following a broadcast application to an agricultural field is an example of nonpoint source pollution.

Distinguish between point sources and nonpoint sources of environmental contamination by pesticides.

Factors That Influence Offsite Movement

There are numerous factors that make offsite movement of pesticides more or less possible in the environment. These include an application site's physical characteristics, the weather, the type of application equipment used, the chemical properties of the pesticide applied, and human behavior.

Describe factors that influence offsite movement of pesticides.

Factors Influencing Movement in Water

Factors influencing runoff and erosion rates include slope, vegetative cover, soil characteristics, volume and rate of water moving downslope, irrigation or rainfall amount and intensity, and temperature. These factors influence how much water runs off and how much moves into the soil (infiltration). Certain persistent pesticides bind to soil, which can then be washed away in water. This is called erosion. In addition, a pesticide's solubility in water contributes to the contamination of surface water, since runoff (the movement of pesticides in water flowing over land) can easily carry soluble pesticides away from the application site.

List features of a given site that influence the potential for a pesticide to reach surface water or groundwater.

Factors influencing leaching (the seepage of water containing pesticides through the soil into groundwater, Fig. 5-6) include a pesticide's chemical and physical characteristics, such as having a high solubility rate, low adsorption rate, or high rate of persistence and soil properties such

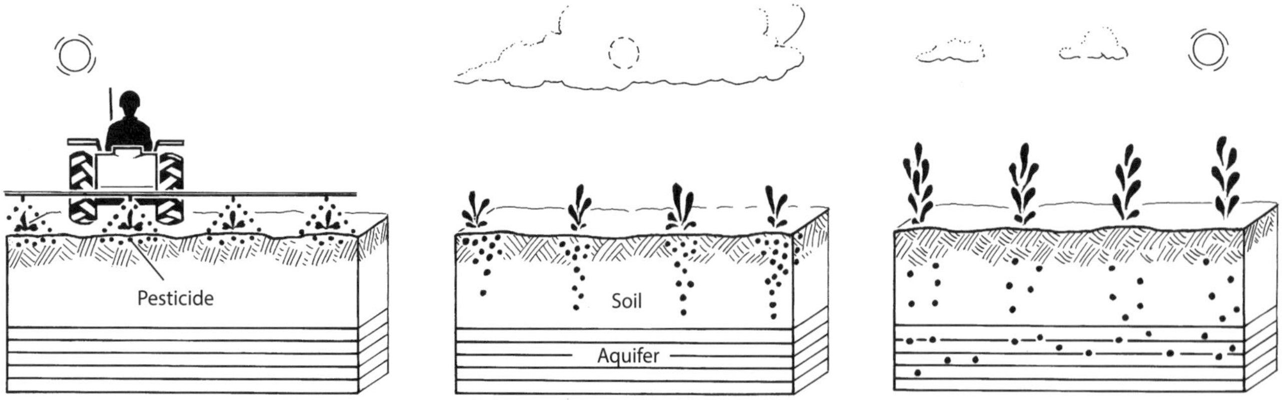

FIGURE 5-6.

Water enters aquifers by percolation through the soil. As water passes down through the soil, it may dissolve some pesticides and carry them into the aquifer. This process is called leaching.

as texture, structure, and amount of organic matter. Figure 5-7 shows a sample pesticide label statement that describes the amount of leaching you can expect when the material is applied to various soil types.

Factors influencing the occurrence of direct-channel contamination of water (Fig. 5-8) include careless handling of pesticides near wells or surface water sources (lakes, rivers, etc.), improper draining of water used to clean application equipment into water bodies or wells, and failing to use backflow protection when filling pesticide tanks or injecting pesticides into an irrigation system.

Factors Influencing Movement in Air

Pesticide movement away from the application site by wind or air currents is called drift. Pesticides may be carried away from a site in the air as spray droplets, vapors, or solid particles, even on windblown soil particles. Factors that influence pesticide drift include volatility of the pesticide,

Soil Texture Guide for Application Rates

Rates listed for incorporated treatments of Treflan HFP are based on Soil Texture Class (coarse, medium, or fine) and soil organic matter content. A fine textured soil (e.g., clay loam) will require a higher application rate than a coarse textured soil (e.g., loamy sand). In the table below, find the Soil Texture Class (coarse, medium, or fine) corresponding to the Soil Texture to be Treated. Choose the proper rate for each application based on the Soil Texture Class and specific crop Directions for Use. Do not exceed the listed maximum use rates.

Soil Texture Class	Soil Texture to be Treated
Coarse (Light) Soils	Sand, loamy sand, sandy loam
Medium Soils	Loam, silty clay loam[1], silt loam, silt, sandy clay loam[1]
Fine (Heavy) Soils	Clay, clay loam, silty clay loam[1], silty clay, sandy clay, sandy clay loam[1]

[1]Silty clay loam and sandy clay loam soils are transitional soils and may be classified as either medium or fine textured soils. If silty clay loam or sandy clay loam soils are predominantly sand or silt, they are usually classified as medium textured soils. If they are predominantly clay, they are usually classified as fine textured soils.

FIGURE 5-7.

An example of a label statement indicating the pesticide's leaching potential when applied to different soil types.

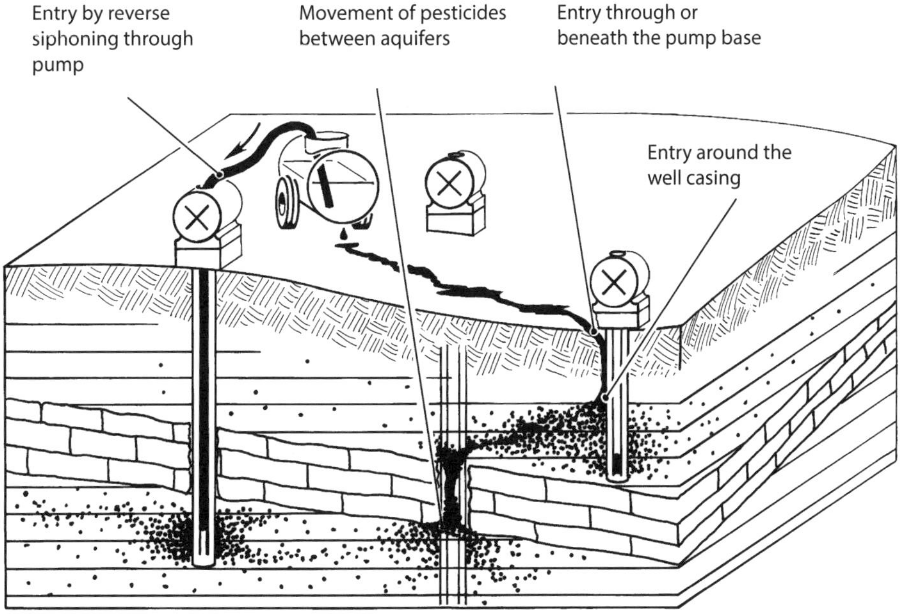

FIGURE 5-8.

Water wells are direct channels into an aquifer and may provide connections between several aquifers. Pesticides and other contaminants can enter groundwater directly through wells.

spray droplet size, release height, wind direction and speed, temperature and humidity, potential for temperature inversions, and rain.

Factors Influencing Movement on or in Objects, Plants, or Animals

Pesticides can move away from the application site on or in objects or organisms that move or are moved offsite. A pesticide's persistence determines its likelihood of moving offsite in this way. For instance, if a pesticide persists in the environment for a long time, it can remain on the surface of plows or other equipment as they are moved around. People or animals who touch that equipment can become sick from accidental exposure. Soil plowed with residue-covered equipment can become contaminated, and that residue may poison beneficial soil-dwelling organisms or kill newly planted crops.

Residues

Explain how pesticide residues can build up on agricultural commodities.

Identify toxicity and residue potential of pesticides that are commonly applied to animals or animal product agricultural commodities.

Whenever you apply a pesticide, it remains as residue on treated surfaces for a time. The chemical nature of the pesticide or the persistence of the formulation affects the amount of residue. The frequency and amount of pesticide used (accumulation) also determine the amount of residue present. For instance, the more often you apply a pesticide to a crop, the more likely you are to experience residue buildup on the surface of those crops. Finally, residues are subject to interaction with the environment (breakdown or recombination).

Residues are important and necessary because they provide the continuous exposure that improves the chances of controlling certain pests, such as parasites that infest livestock. However, residues are undesirable when they expose people, domestic animals, or wildlife to unsafe levels of pesticides. Pesticide materials that miss the treatment surface can remain as residues in soil, water, or nontarget areas. Residues can also be transferred to animal products like milk when pesticides remain on an animal's skin, coat, or hair. Also, empty pesticide containers hold small amounts of residues. Those residues, along with the containers themselves, require proper disposal to prevent environmental contamination (Fig. 5-9).

Avoiding Hazardous Residues

Reduce chances of creating hazardous pesticide residues by taking the following steps:

- Comply with label restrictions on timing, placement, and rate of application, using the lowest effective rate whenever possible.
- Perform compatibility tests before tank-mixing two or more pesticides.
- Apply pesticides during dormant or fallow periods, or when cows, goats, or sheep are not lactating, to prevent spraying edible produce or contaminating animal products whenever possible.
- Avoid pesticide spills and clean up an accidental spill immediately.

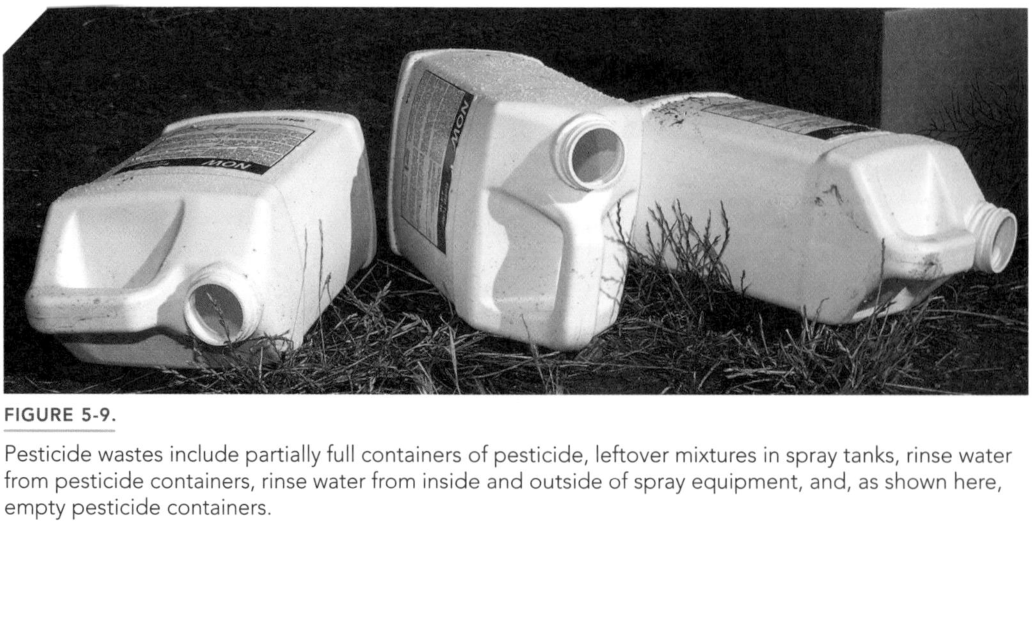

FIGURE 5-9.

Pesticide wastes include partially full containers of pesticide, leftover mixtures in spray tanks, rinse water from pesticide containers, rinse water from inside and outside of spray equipment, and, as shown here, empty pesticide containers.

- Fill application equipment using an air gap or check valve to prevent pesticide mixtures from siphoning back into wells.
- Calibrate application equipment properly, and make an accurate measurement of the area you plan to spray.
- Select pesticides that break down rapidly when possible, and use formulations that reduce the likelihood of drift.
- Control the amount and timing of irrigation water to eliminate runoff and slow the rate of seepage.
- Reduce soil erosion by using practices such as reduced tillage, contour farming, terracing, grass-lined waterways, and subsurface drainage.
- Collect and reuse tailwater (the water that runs off the low end of a field) from irrigated fields to keep residues within the treatment site.

Environmental Impacts of Pesticide Application

Some types of pesticide you use may be harmful to nontarget organisms at the application site and in the surrounding environment, including nearby sensitive areas such as rivers, lakes, and aquifers. Before making a pesticide application, become familiar with the treatment area and its surroundings. Avoid using pesticides that disrupt natural enemies and other beneficial organisms, are likely to make their way into freshwater supplies, or harm wildlife or nontarget plants.

Remember that the result of unintended pesticide exposure is not always immediately apparent. Long-term consequences of offsite movement may include accumulation of pesticides in animals and soil, problems with wildlife breeding and offspring, and the development of diseases in otherwise healthy organisms long after the initial exposure.

Impacts on Water Supplies and Sensitive Areas

Groundwater. Potential contamination of groundwater with pesticides is a serious concern, because we rely on it for drinking, irrigation, and many other purposes. About two-thirds of the water we rely on is groundwater—it is our most important source of freshwater.

Groundwater in California is especially vulnerable to contamination because it is directly below so much cultivated, industrial, and residential land. Contamination, when it occurs, may be very hard or impossible to contain. You must be especially careful to keep pesticides out of groundwater, because it can take hundreds of years to remove those contaminants.

Surface Water. Surface water (in ditches, streams, rivers, ponds, and lakes) is also an important source of drinking water. Therefore, pesticide contamination of surface water is a health concern. Pesticides that move in runoff water or with eroded sediment may reach sources of surface water, creating a health hazard for people and wildlife in the area.

Sensitive Areas. In addition to water sources, sensitive areas include sites where living things could easily be injured by a pesticide. Sensitive areas include (but are not limited to)

- schools, playgrounds, recreational areas, hospitals, and neighborhoods
- habitats of endangered species
- apiaries and honey bee habitat, wildlife refuges, and parks
- places where domestic animals and livestock are kept, confined, or cared for
- fields of food or feed crops not listed on the pesticide label

Whenever possible, take special precautions to avoid application to the sensitive area. Leaving an untreated area around a sensitive site is a practical way to avoid contaminating it. In still other instances, the sensitive area may be near a site used for pesticide mixing and loading, storage, disposal, or equipment washing. You must take precautions or relocate your work site to avoid accidental contamination of the sensitive area. Check the label for statements that alert you to special restrictions around sensitive areas.

California has strict regulations around pesticide applications made to agricultural commodities produced near public K-12 schools and licensed day care facilities (collectively referred to as "school sites"). In addition to requiring notification to the school site of expected pesticide applications, the regulations establish pesticide application restrictions Monday through Friday, from 6:00 a.m. to 6:00 p.m. (except on holidays and during school breaks when facilities are closed all day), within a specified distance of a school site. Two types of restriction distances for outdoor applications apply: either ¼ mile (1,320 feet) or 25 feet, depending on the type of application equipment used and type of pesticide applied. There are no distance restrictions when the application is made within an enclosed space, unless a fumigant is applied, which is prohibited within ¼ mile of a school site. Additionally, backpack sprayers that incorporate an air blast sprayer are prohibited within ¼ mile of a school site.

Impacts on Nontarget Organisms

Describe ways that pesticides can impact nontarget organisms.

Pesticides may affect nontarget organisms directly, causing immediate injury. When a pesticide is nonselective, it will kill the pest as well as many of the pest's natural enemies. Sometimes when you apply a more selective chemical, your application may still affect natural enemies by destroying the pest they depend on for food. After natural enemies die off or leave because of the lack of prey, they often require more time than the pest to increase their population size. Because the area lacks natural enemies that normally keep the pests in check, pest populations can grow rapidly and sometimes become bigger than before the pesticide treatment. This phenomenon is known as pest resurgence.

Endangered Species. An endangered species is on the brink of extinction throughout all or a significant portion of its range. A threatened species is likely to become endangered in the near future. Enforcement agencies restrict the use of certain pesticides in areas where endangered species live. For more information, consult your local University of California Cooperative Extension office. Advisors in these offices can also give you information on nonchemical pest control methods and help you integrate these into existing pest management plans. For more information on how to create an IPM program, see Chapters 1 and 11. For information on endangered species laws, identification of areas of endangered species habitat, and ways to protect endangered species, check with the local or regional office of the California Department of Fish and Wildlife, your local county agricultural commissioner, or the DPR website. DPR maintains an online database called PRESCRIBE (cdpr.ca.gov/docs/endspec/prescint.htm) that can help you find out if there are any endangered species living near your application site. The database also lists the restrictions in effect for pesticides that you may want to apply at that site.

Bees and Other Pollinators. Certain types of insecticide and fungicide applications may kill honey bees and other pollinators. Pollinators are most susceptible if you apply harmful pesticides while they are foraging for nectar and pollen. Certain pesticides may hurt social bees that live in hives if they are brought there by the foraging adults. Because of the importance of pollinators in our environment, you must notify beekeepers within 1 mile of the application site if the pesticide you are applying may injure bees and if the plants that bees will visit are in bloom. Sidebar 5-1 provides an example of the "Protection of Pollinators" statement from a pesticide label for a product that is known to be hazardous to bees.

Other Impacts. Applications of persistent pesticides can lead to secondary poisoning of nontarget organisms (bioaccumulation). Bioaccumulation (Fig. 5-10) occurs when certain pesticides slowly build up within the bodies of predators that feed on animals that have eaten or absorbed small amounts of these pesticides. Greater amounts of these pesticides build up in the bodies of predators as time passes, which may affect their health and ability to reproduce and can cause premature death. Predators can also be killed after exposure to a large dose of pesticide, which happens when they feed on pesticide-poisoned rodents. These unintended pesticide-related deaths are called secondary kills.

Pesticides applied to livestock and poultry can hurt the animals if they accidently ingest the pesticide (when grooming, for instance) or if pesticides are overapplied. Symptoms of pesticide poisoning in farm animals are similar to those observed in humans. Call a veterinarian immediately if you notice skin irritation, discomfort or blistering; excessive salivation, tremors, vomiting, depression, or hyperexcitability; or fever, breathing trouble, disorientation, seizures, or death. Make sure the pesticide label and SDS are readily available to help with diagnosis and treatment.

If you improperly apply herbicides, you may accidentally kill nontarget plants, including nearby crops. Many species of plants are important in natural and undeveloped areas on the farm where you work because they protect the watershed, reduce erosion, provide food and shelter to beneficial organisms and wildlife, and are part of the natural flora. When the ecological balance of an area is disrupted, such as with the unintentional destruction of natural flora by herbicides, weedy plants are likely to take over. These undesirable species usually fail to provide the natural food and shelter needed by beneficial organisms and wildlife.

Another problem associated with pesticide use is that of secondary pest outbreak. Secondary pests are normally controlled by natural enemies or competition from the primary pest. Eliminating natural enemies or primary pests often results in an increase in secondary pest populations that can cause economic damage.

SIDEBAR 5-1

Sample U.S. EPA Bee Label, "Protection of Pollinators"

PROTECTION OF POLLINATORS

APPLICATION RESTRICTIONS EXIST FOR THIS PRODUCT BECAUSE OF RISK TO BEES AND OTHER INSECT POLLINATORS. FOLLOW APPLICATION RESTRICTIONS FOUND IN THE DIRECTIONS FOR USE TO PROTECT POLLINATORS.

Look for the bee hazard icon  in the Directions for Use for each application site for specific use restrictions and instructions to protect bees and other insect pollinators.

This product can kill bees and other insect pollinators.

Bees and other insect pollinators will forage on plants when they flower, shed pollen, or produce nectar.

Bees and other insect pollinators can be exposed to this pesticide from:

- Direct contact during foliar applications, or contact with residues on plant surfaces after foliar applications
- Ingestion of residues in nectar and pollen when the pesticide is applied as a seed treatment, soil, tree injection, as well as foliar applications.

When Using This Product Take Steps To:

- Minimize exposure of this product to bees and other insect pollinators when they are foraging on pollinator attractive plants around the application site.
- Minimize drift of this product on to beehives or to off-site pollinator attractive habitat. Drift of this product onto beehives or off-site to pollinator attractive habitat can result in bee kills.

Information on protecting bees and other insect pollinators may be found at the Pesticide Environmental Stewardship website at: http://pesticidestewardship.org/PollinatorProtection/Pages/default.aspx.

Pesticide incidents (for example, bee kills) should immediately be reported to the state/tribal lead agency. For contact information for your state, go to: www.aapco.org/officials.html. Pesticide incidents should also be reported to the National Pesticide Information Center at: www.npic.orst.edu or directly to EPA at: beekill@epa.gov

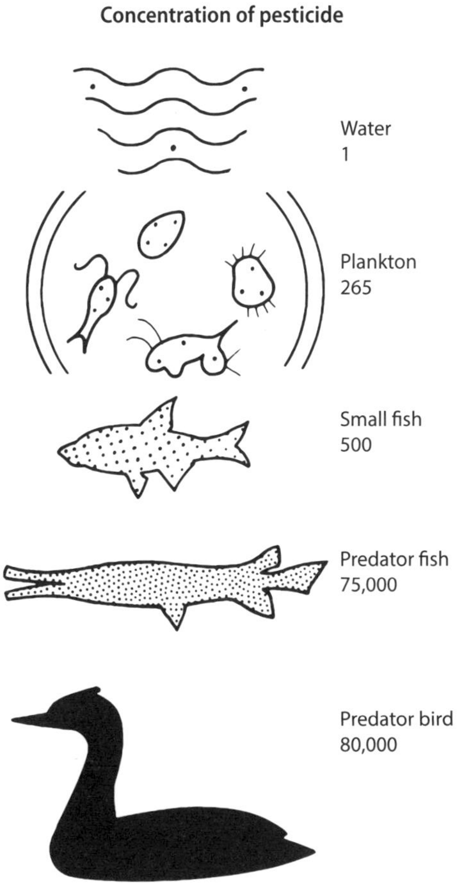

FIGURE 5-10.

Bioaccumulation is the way pesticides become concentrated through the biological food chain. Microorganisms and algae containing pesticides are eaten by small invertebrates and hatching fish. Larger fish and birds eat these organisms in turn. Each passes greater amounts of pesticide to the larger animal.

Chapter 5 Review Questions

1. True or false?

☐ True ☐ False a. Groundwater contamination is a problem when using persistent pesticides.

☐ True ☐ False b. Point source pollution comes from pesticides that have been spilled over a wide area.

☐ True ☐ False c. Pesticides that drift from the target site can injure nontarget organisms.

☐ True ☐ False d. Noting where sensitive areas are on your farm before applying pesticides makes it more likely you will contaminate these areas.

☐ True ☐ False e. Nonpoint source pollution comes from pesticides that move into streams or groundwater following a broadcast application to a large area.

2. Match the term with its definition.

1. solubility	a. the ability of a pesticide to remain present and active in its original form for an extended period before breaking down
2. adsorption	b. the tendency of a pesticide to turn into a gas or vapor
3. persistence	c. a measure of the ability of a pesticide to dissolve in liquid
4. volatility	d. the process a pesticide undergoes when it binds to soil particles

3. Match the situation with the type of offsite movement most likely to occur.

1. A pesticide is being applied at high pressure using small-orifice nozzles, and the wind picks up speed.	a. drift
2. Rainwater washes a soluble pesticide down through the soil into an aquifer.	
3. Irrigation water carries pesticides from a recently treated field into a nearby creek.	b. runoff
4. A pesticide is applied during the formation of a temperature inversion.	
5. A soluble pesticide is applied on a slope just before a rainstorm.	c. leaching
6. Rinse water from equipment cleaning is dumped next to a treated field in a Ground Water Protection Area.	

4. The more often you apply a pesticide to a crop, the more likely you are to experience ________.

☐ a. buildup of residues on crop surfaces
☐ b. growth of crops at faster rates
☐ c. concentration of invasive species

5. Bioaccumulation can happen to predators in the environment when they repeatedly ______________.

☐ a. groom themselves after entering a recently treated area
☐ b. eat organisms that were exposed to certain pesticides
☐ c. carry pesticide residues back to burrows, dens, or nests

6. A nonselective insecticide can kill ________.

☐ a. the insect pest and its natural enemies
☐ b. only the secondary insect pests
☐ c. a single type of insect pest

CHAPTER 6
HUMAN HAZARDS

Potential for Human Injury 68
Harmful Effects of Pesticide Exposure 72
Other Problems Associated with Pesticides.......................... 74
Chapter 6 Review Questions 75

Knowledge Expectations

1. Describe the ways people get exposed to pesticides and the routes of entry.
2. List the tasks most often associated with accidental pesticide exposure and explain why these tasks are hazardous.
3. Describe how offsite movement of pesticides endangers human health.
4. Name conditions at the application site that may change and influence the hazards associated with pesticide application.
5. Explain the human hazards associated with pesticides.
6. List the hazards associated with pesticides commonly used on or around animals.
7. Explain how each of the following can contribute to human hazards associated with pesticide use:
    a. incorrect dosage
    b. incorrect application timing
    c. incorrect pesticide product application
8. Describe the potential effects of pesticide exposure on people (acute, chronic).

Potential for Human Injury

Pesticides, like other poisonous chemicals, injure people by interfering with biological functions. The type and degree of injury depend on the pesticide's toxicity and the amount entering the tissues. Some pesticides are very toxic and produce injury at low doses. A few drops of these might cause severe illness or death.

Potential hazards exist with all pesticides, even the least toxic products, so you must avoid exposure when working with them. Treat all pesticides with respect. It is impossible to accurately predict what effects can result from long-term, repeated exposures to pesticides.

Pesticide poisoning symptoms can vary widely. If you suspect you have been exposed to pesticides, seek medical attention and be prepared to describe the pesticide and how you might have been exposed to it.

How People Get Exposed to Pesticides

Describe the ways people get exposed to pesticides and the routes of entry.

List the tasks most often associated with accidental pesticide exposure and explain why these tasks are hazardous.

There are several ways people come in contact with pesticides, including during mixing and loading, when applying, and even during the laundering of work clothing. The most serious exposure incidents happen when pesticide containers are mishandled during the transportation and storage of pesticides. Transportation and storage activities are particularly dangerous because the pesticides you handle are concentrated. When accidents happen, exposure to these concentrated chemicals can cause more harm to people and the environment. Mixing and application are also frequent causes of injury due to pesticide overexposure. To greatly reduce exposure risks, secure pesticides correctly for transport, store pesticides in proper containers and in properly equipped facilities, wear proper work clothing, and use required personal protective equipment (PPE) during all handling tasks. In addition, following label requirements for restricted-entry and preharvest intervals and the guidelines for cleaning and storing PPE helps protect everyone.

As you work with pesticides, accidental spills may result in serious exposure. Protective clothing and prompt emergency response reduce the chances of serious injury if you have an accident.

Describe how offsite movement of pesticides endangers human health.

It is also possible for people to be exposed to small amounts of pesticide if they live near areas where you are spraying. For instance, drift events can result in pesticide residues above legal limits on fruits and vegetables harvested from people's backyard gardens. It can also contaminate laundry hanging on lines near application sites, land on toys left outside in nearby yards, or enter homes through open windows. Runoff and leaching can contaminate people's drinking water by polluting wells, lakes, streams, and aquifers. You must think about the location of people and the resources they rely on as you plan your application. Even when pesticides stay on target, people can be exposed to pesticide residue if they eat treated produce before the harvest interval expires or touch recently treated plant foliage. Careful planning can help you avoid accidentally exposing yourself and others to pesticides.

One of the most tragic types of pesticide injury is caused by storing pesticides in food or drink containers (Fig. 6-1). Many cases have been reported of children drinking pesticides from soft drink containers. Never store pesticides in anything other than the containers in which they were purchased. Don't take farm chemicals

FIGURE 6-1.

Children are the major group of nonagricultural pesticide poisoning victims. Improper storage of pesticides brought home from work sites is one of the primary ways children find and ingest pesticides.

home to use around the house. Unless you have control over the containers, keep pesticides locked up in a storage area that is inaccessible to children and untrained adults.

Work-Related Exposure

Name conditions at the application site that may change and influence the hazards associated with pesticide application.

Explain the human hazards associated with pesticides.

List the hazards associated with pesticides commonly used on or around animals.

Pesticide applicators and handlers are most at risk from pesticide exposure because they work closely with toxic materials. Applicators are most at risk during mixing and loading because of the increased chance of exposure from spills and splashes of concentrated pesticides. Exposure to concentrated pesticides is also what makes transportation and storage of pesticides so hazardous to people. Laborers, tractor drivers, irrigators, and other farm employees risk exposure if they transport pesticides or work in recently treated areas. Restricted-entry intervals are important pesticide use restrictions designed to protect agricultural workers from exposure (Fig. 6-2). Methods such as noting conditions at the application site that can change quickly and increase drift potential (like wind speed and direction, temperature, cloud cover, etc.) and making spray applications when workers are not present nearby also help. Another important step is to train employees on how to avoid contact with pesticide residues. For more information about variables that can affect hazards at application sites, see Chapters 5 and 8.

People who maintain or repair application equipment may contact pesticide residues on that equipment. Oil-soluble pesticides are a major concern. These build up in grease deposits and on oily surfaces and may be hard to remove. Cleaning the application equipment often helps reduce pesticide residue and lowers risks to maintenance workers and operators. If you cannot clean equipment before repairs or maintenance, mechanics must wear required PPE to avoid exposure. People who clean or repair pesticide-contaminated equipment are considered pesticide handlers and must receive pesticide handler training.

FIGURE 6-2.

Restricted-entry intervals following agricultural pesticide applications have helped to reduce farmworker injury. Growers often post treated fields, like the one shown here, to warn workers not to enter without PPE during the restricted-entry interval.

When you use persistent pesticides, accidental exposure can occur even after harvest. To protect consumers and farmworkers from pesticide exposure, regulations establish preharvest intervals (the fewest number of days before harvest that a pesticide can be applied) for treated produce.

It is hard for greenhouse and nursery workers to avoid close contact with treated surfaces because greenhouses often have lots of plants in a limited space. Also, most greenhouses have limited ventilation, which can increase the potential for breathing spray mists or vapors. It also raises the risk of getting dusts or mists onto the skin or into the eyes during applications.

If any of the farm's employees handle pesticides applied around or to livestock or poultry, they must understand the task's unique hazards and how to avoid them. These hazards include increased chances of exposure via splashing liquids from dipping vats for sheep and cattle and airborne dusts released from dust bags for livestock or dust boxes for poultry. Also, because some sprays are used in enclosed spaces such as henhouses, workers in animal agriculture will have some of the same problems as those who work in greenhouses or other spaces with limited ventilation.

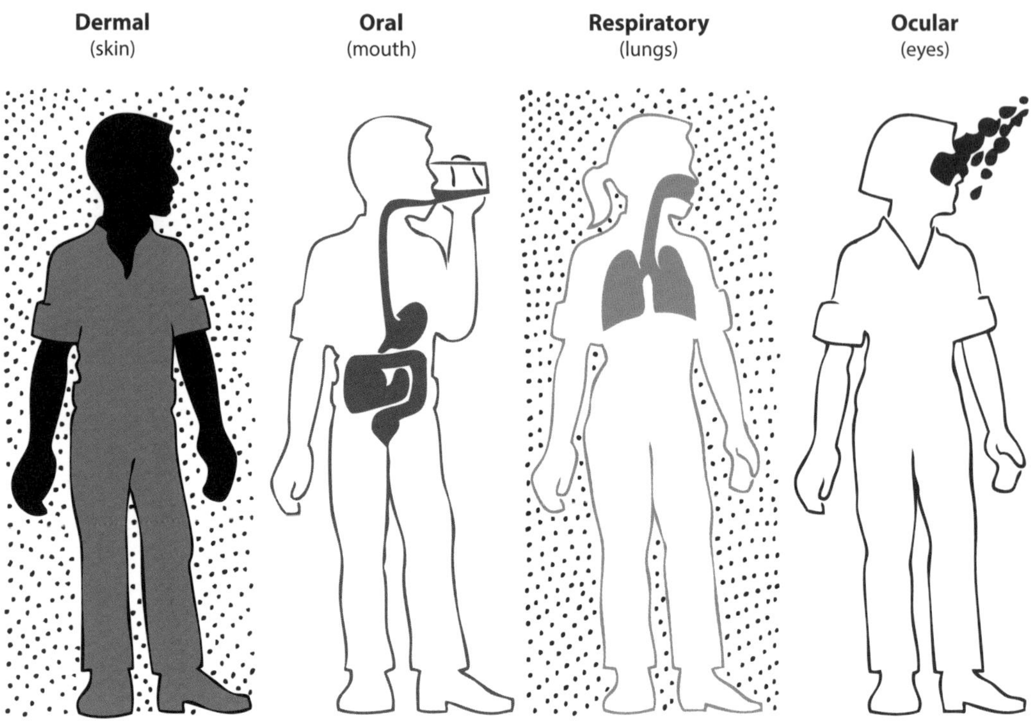

FIGURE 6-3.

The most common ways for pesticide exposure to occur are through the skin (dermal), through the mouth (oral), through the lungs (respiratory), and through the eyes (ocular).

How Pesticides Enter the Body

The tissues of an exposed person can absorb certain types of pesticides. These pesticides enter the body through the skin, eyes, lungs, or mouth (Fig. 6-3). You can find detailed emergency response and treatment methods for all exposure types in Chapter 12.

Skin Exposure

Skin (or dermal) contact is the most common route of pesticide exposure. If certain pesticides contact the skin, they may cause a rash or mild irritation (known as dermatitis). Other types of pesticides cause more severe injury, such as burns. You can also be poisoned if your skin absorbs a pesticide. When pesticides are absorbed through your skin, blood will carry them to other organs in the body.

The ability of a pesticide to pass through your skin depends on its chemical characteristics and formulation. Oil-soluble pesticides pass through skin more easily than those that dissolve in water, for instance. The amount of pesticide absorbed by your skin also depends on which body parts are exposed. In a test using the organophosphate insecticide parathion (Fig. 6-4), researchers found the forearm to be the least susceptible area for pesticide absorption. The palms of the hands absorb parathion slightly faster than the forearm. The scalp, face, and forehead are four times more susceptible. In this study, the genital area was the most susceptible area

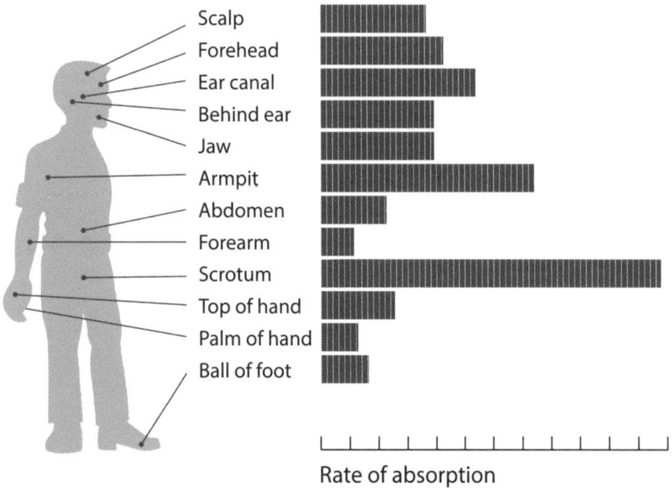

FIGURE 6-4.

Different areas of the body absorb pesticides through the skin at different rates. This illustration shows the results of an early study in which researchers placed minute amounts of methyl parathion on body areas of various volunteers. They determined absorption rates by measuring the chemical in the volunteers' urine after a known period of time. (Testing pesticides on human subjects is not allowed in the United States.) *Source:* Feldmann 1967.

of the body to parathion absorption. This area was nearly 12 times more susceptible than the forearm.

To prevent skin exposure to pesticides, always wear work clothing and the required PPE when handling is part of your job or when you will be entering recently treated sites. Be sure to wash your hands thoroughly before using the restroom when working with or around pesticides. Also, avoid contact with recently treated plants, animals, and commodities whenever possible. See Chapter 7 for more about PPE.

Eye Exposure

Some pesticide formulations can hurt your eyes. In addition, the eyes provide another route for entry of certain pesticides into your body. California law requires that protective eyewear be worn

- during all mixing and loading activities
- while adjusting, cleaning, or repairing contaminated mixing, loading, or application equipment
- during most types of ground application

Protect your eyes by wearing a face shield, goggles, or safety glasses.

Respiratory Exposure

The lungs quickly absorb certain pesticides, and the blood transports these pesticides to other parts of the body. Some pesticides cause serious lung injury. Breathing dust or vapor during mixing or application is hard to avoid unless one uses appropriate respiratory equipment. Always wear label-required respirators during mixing and application. If pesticide labels, regulations, or employer policies mandate respirator use, employee handlers must be evaluated by a doctor and declared medically fit before they can wear a respirator on the job. Make sure any respirator you use fits properly and is in good condition. Fit testing must be performed prior to your first use of a respirator and annually thereafter (Fig. 6-5). In addition to this yearly testing, you must fit check your respirator before each use to be sure it will protect you while you are handling pesticides. For more information about respirators, fit testing, and fit checking, see Chapter 7. For a detailed list of requirements for respirator use by employee handlers, see the Pesticide Safety Information Series A-5.

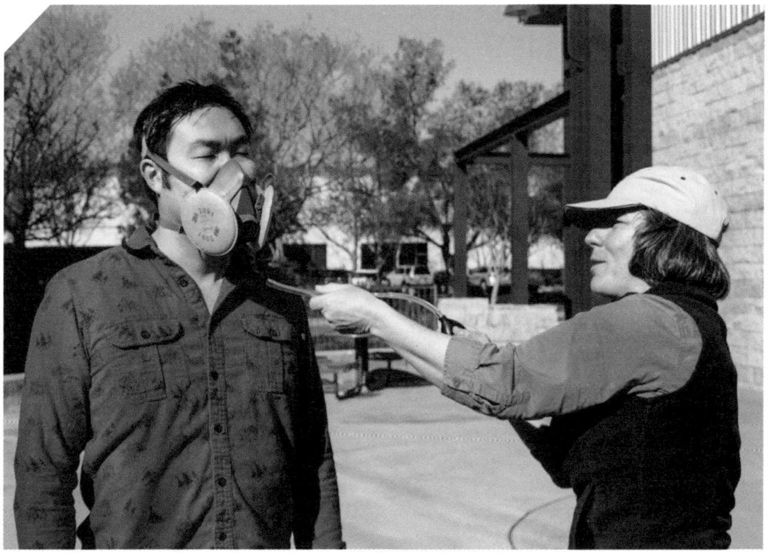

FIGURE 6-5.

A specially trained staff person from UC ANR's Environmental Health and Safety department performs a respirator fit test for a visiting pesticide applicator.

Exposure through the Mouth (Oral)

It is very rare for someone to accidentally drink or eat a pesticide. The exceptions are when pesticides are improperly stored or put into food or drink containers. Pesticides stored in food or drink containers can easily be mistaken for something safe to eat or drink. Improperly stored pesticides are especially dangerous if children will be nearby.

Exposure through the mouth occurs more commonly if spray materials or pesticide dusts splash or blow into your mouth during mixing or application. Sometimes swallowing a pesticide happens when you eat or drink something that has been contaminated. You may also swallow pesticides if you smoke or put gum in your mouth while handling pesticides.

Linings of your mouth, stomach, and intestines readily absorb some pesticides. If you swallow enough, sometimes even small amounts, you may get sick. PPE, such as a respirator or face shield, lowers the risk of pesticides getting into your mouth. See Chapter 7 for PPE that helps protect people from oral exposure.

Before eating, drinking, or smoking, be sure to wash your hands thoroughly. Keep food and drinks away from areas where pesticides are being applied or mixed. Never put pesticides into food or drink containers. Keep all pesticides in their original packages. Do not mix or measure pesticides with utensils that someone could use later to make or serve food, and do not use food containers to measure pesticides.

How Misapplication of Pesticides Endangers Human Health

Explain how each of the following can contribute to human hazards associated with pesticide use:

- incorrect dosage
- incorrect application timing
- incorrect pesticide product application

Pesticide misapplication can occur at any time and may be the result of intentional, accidental, or careless behavior by the pesticide handler. Misapplication of pesticides is dangerous for many reasons, and you should take care when applying pesticides to avoid situations that contribute to errors before, during, and after pesticide handling activities.

Misapplications include applying the wrong pesticide, applying the wrong amount of pesticide, and applying pesticide to the wrong site or crop.

Applying the Wrong Pesticide

Lack of attention to your mixing operation or giving the wrong instructions to an employee may result in the wrong pesticide being applied. Besides possible damage to plants or surfaces in the treatment area, using the wrong pesticide exposes you and other employees to additional hazards. Mixing and application might take place while using the wrong personal protective equipment, resulting in possible injury to the applicator.

Applying the Wrong Amount of Pesticide

There are two types of mistakes people make when they apply the wrong amount of pesticide:

1. They apply too little of the pesticide.
2. They apply too much of the pesticide.

Applying less than the label rate of a pesticide isn't illegal, but it often fails to control the target pest, wasting time and money. The most important problem caused by applying less than labeled rates, however, is the development of pesticide resistance. If pesticide resistance is allowed to progress, a pesticide that previously worked when applied at label rates no longer does so. Pests that survive repeated applications of the same pesticide require the introduction of new pesticides that take time and money to develop, and that may be more hazardous to people and the environment.

Applying more than the label rate of a pesticide at a site is illegal—it can damage the environment and threaten human health. This type of problem occurs as a result of poorly calibrated equipment, incorrect mixing of chemicals in your spray tank, or miscalculation of the label's application rates. Residues from an overapplied pesticide may last longer than expected or cause damage to surfaces in the treated area.

Applying a Pesticide to the Wrong Site or Crop

Another form of accident involves applying pesticides to the wrong site. This can be a serious problem if the site (or crop) is not listed on the pesticide label or if there are people working at the site.

Harmful Effects of Pesticide Exposure

Describe the potential effects of pesticide exposure on people (acute, chronic).

Acute Effects of Pesticide Exposure

Acute toxicity is the measure of harm (systemic or contact) caused by a single, one-time exposure event. Acute effects of pesticide exposure occur shortly after the event, usually within 24 hours. The symptoms from an acute pesticide exposure event depend on the dose, the exposure period, and your own body chemistry and weight.

Manufacturers list systemic and contact effects in addition to the signal word on the label. Systemic and contact acute toxicity are indicated by the signal words and further explained in the "Precautionary Statements" portion of the product label in the "Hazards to Humans and Domestic Animals" section.

Chronic Effects of Pesticide Exposure

The chronic toxicity of a pesticide is determined by subjecting test animals to long-term exposure to an active ingredient, typically 2 years. The harmful effects that occur from small, repeated doses over time are called chronic effects. If a product causes chronic effects in laboratory animals, the manufacturer is required to include chronic toxicity warning statements on the product label. This information is also listed on the Safety Data Sheet. The chronic toxicity of a pesticide is harder to determine through laboratory analysis than the acute toxicity. Remember that signal words for specific formulations are determined by using data from acute toxicity studies and do not take into account chronic, or long-term, effects.

Sensitization

Sensitization is the gradual development of an allergic reaction to a type of pesticide or chemical. It is similar to what happens to some people after they touch poison oak repeatedly over time. The first several exposures to poison oak may cause a mild rash or may not cause any problems at all. However, over time people can become sensitized to the plant. Once people become sensitized, they will develop a rash that gets worse every time they touch poison oak.

Likewise, if you are exposed to a normal amount of a particular pesticide again and again, you can become sensitized to it. Once you have become sensitized to a pesticide, as with poison oak, symptoms are likely to worsen with repeated contact. Normal exposure is not the same thing as overexposure to pesticides. If you get headaches, rashes, or become dizzy when working with a particular pesticide or when you enter an area where the pesticide was recently used, you may have become sensitized to it. Not everyone will become sensitized to pesticides, but those who do should avoid exposure to the pesticide causing the bad reaction.

Delayed Effects of Pesticide Exposure

Delayed effects are illnesses or injuries that do not appear immediately (within 24 hours) after exposure to a pesticide. They may be delayed for weeks, months, or even years. Whether you experience delayed effects depends on the pesticide, the extent and route of exposure, and how often you were exposed. Under "Precautionary Statements," the label states any delayed effects that the pesticide might cause and how to avoid exposures leading to them. Wearing extra protective gear and taking additional precautions may be necessary to reduce the risk of delayed effects. Delayed effects may be caused by either an acute exposure or chronic exposure to a pesticide.

Symptoms of Pesticide Exposure

You can be injured either by a single massive dose being absorbed during one pesticide exposure (acute) or from smaller doses absorbed during repeated exposures over an extended period of time (chronic). Accidental exposure to some pesticides may cause irreversible or permanent damage, which can result in long-term illness, disability, or death. Moderate exposure incidents can result in more general symptoms such as sweating, nausea, chest pains, skin irritation, swelling, or dizziness, among others.

When you have been exposed to a large enough dose of pesticide to produce injury or poisoning, you may experience either an immediate or delayed appearance of symptoms. Immediate symptoms are those observed soon after exposure—known as acute onset. Delayed symptoms are those that appear 24–48 hours after the exposure incident. Sometimes symptoms from pesticide exposure may not show up for weeks, months, or even years. These are called chronic symptoms and can easily be mistaken for other types of illness, since they can appear long after the initial exposure event.

Other Problems Associated with Pesticides

Heat-Related Illness. When you apply pesticides wearing any type of PPE, you may end up with symptoms of heat-related illness. Heat-related illness happens when your body cannot cool down, which causes your core temperature to rise. This condition can occur when the air temperature is close to or warmer than normal body temperature and humidity is high. Blood circulated to your skin cannot lose its heat, and so you begin to sweat as a way to cool off. But sweating is effective only if the humidity level is low enough to allow evaporation and if the fluids and salts that are lost are replaced often enough.

If your body cannot get rid of excess heat, it will store it. When this happens, your body's core temperature rises and your heart rate speeds up. As your body continues to store heat, you begin to lose concentration and have a hard time focusing on work, may become irritable or sick, and often lose the desire to drink. The next stage is most often fainting. Heat-related illness can even lead to death if you are not cooled down.

Heat-related illness symptoms may look like certain types of pesticide poisoning, so provide emergency or medical personnel with detailed information about the events surrounding the incident. California regulations require that employee handlers and fieldworkers receive training on recognizing, avoiding, and treating heat-related illness along with training on recognizing pesticide illness. For more information, see California OSHA regulations, section 3395, "Heat Illness Prevention."

Chapter 6 Review Questions

1. The most common route of pesticide exposure is through the _______.

- ☐ a. mouth
- ☐ b. skin
- ☐ c. eye

2. Offsite movement of pesticides can endanger human health in which of these ways? Select all that apply.

- ☐ a. It can cause residues to exceed legal limits on food crops.
- ☐ b. It can contaminate fruits and vegetables in backyard gardens.
- ☐ c. It can pollute surface water and groundwater used for drinking.
- ☐ d. It can contaminate laundry hung outside to dry.

3. Which conditions at a site can change quickly and affect the outcome of your pesticide application? Select all that apply.

- ☐ a. soil type and contents
- ☐ b. wind speed and direction
- ☐ c. temperature and cloud cover
- ☐ d. presence of lakes and streams

4. Mixing and loading are considered among the most risky activities because __________.

- ☐ a. recommended PPE is not protective enough when people are in close contact with concentrated pesticides
- ☐ b. label directions are often hard to follow when measuring and mixing concentrated pesticides
- ☐ c. spills and splashes are more dangerous when working with concentrated pesticides

5. What hazards does the preharvest interval help you avoid? Select all that apply.

- ☐ a. exposing consumers to unsafe levels of pesticide residue on fruits and vegetables
- ☐ b. exposing fieldworkers to excessive pesticide residues on crops they harvest
- ☐ c. exposing yourself to dangerously high pesticide residues during application

6. What is the difference between chronic and acute pesticide exposure?

- ☐ a. Chronic exposure is the result of short-term contact with any amount of pesticide; acute exposure is a result of long-term contact with a small amount of pesticide.
- ☐ b. Chronic exposure results from a single low-dose incident; acute exposure results from a single high-dose incident.
- ☐ c. Chronic exposure is repeated exposures to small amounts of pesticide; acute exposure is a short-term exposure to a large dose of pesticide.

7. On a humid summer day, you notice that a co-worker has trouble focusing on the job he is doing, is irritable, and starts to complain of feeling sick. When you offer him a cool drink, he shows no interest in it. Your co-worker is suffering from ______________.

- ☐ a. heat-related illness
- ☐ b. exhaustion
- ☐ c. the flu

Chapter 7
Personal Protective Equipment and Personal Safety

Keeping People Safe 78
Personal Protective Equipment 80
Engineering Controls 90
Chapter 7 Review Questions 93

Knowledge Expectations

1. Describe safety training and recordkeeping for fieldworkers and pesticide handlers.
2. Describe the farm owner or manager's responsibility to provide personal protective equipment (PPE) for mixing, loading, applying, and storing pesticides to employees.
3. Explain how PPE can protect a person from hazards associated with pesticides.
4. List various PPE that pesticide handlers use to protect themselves from pesticide exposure.
5. Explain how to select the most effective PPE for the job, including understanding its limits to protect you.
6. Describe how to wear, clean, maintain, and store reusable PPE, and how to dispose of worn or single-use PPE.
7. List the different kinds of engineering controls and explain when these are used.

Keeping People Safe

Describe safety training and recordkeeping for fieldworkers and pesticide handlers.

All pesticide handlers—applicators, mixers, loaders, flaggers, etc.—and early-entry agricultural workers (those who enter an area before the restricted-entry interval has ended) are legally required to receive safety training and to follow all PPE instructions on the product label and in California laws and regulations. A pesticide label lists the minimum PPE that you must wear while doing handling or early-entry activities, and California regulations often require more than what is listed on the label. You may decide to wear additional PPE for increased safety beyond what is required by regulation; however, you may not leave out any equipment mentioned on the label or in regulation, and you must follow the most restrictive PPE requirements listed. Sometimes a label has different PPE requirements for pesticide handlers and early-entry workers, so read the label carefully (see Chapter 4 for how to find PPE requirements on pesticide labels).

The following sections describe required training for employee handlers and fieldworkers. They also define the types of PPE typically available to protect you from pesticide exposure and discuss how to clean and maintain your PPE so you stay safe on the job.

SIDEBAR 7-1

Criteria for Pesticide Handler Training

Annual pesticide worker safety training for employee handlers must include at least the following information. Training is not required if workers or employers can verify such training has been provided within the past year.

ADDITIONAL INFORMATION THAT MUST BE COVERED

- How to handle, open, and lift containers; how to pour; and how to operate mixing and application equipment
- How to properly triple-rinse and dispose of containers
- How to confine the pesticide to the application area or site
- How to avoid contamination of people, animals, waterways, and sensitive areas
- How to handle nonroutine tasks or emergency situations such as spills, leaks, or fires
- How and where to store containers; how to proceed when containers cannot be locked up
- How to read and understand pesticide labels and Safety Data Sheets, including the signal words, precautionary statements, first aid instructions, application rate, and mixing and application instructions
- Why and when to wear different types of PPE
- How to fit and properly wear PPE, and how to inspect it for wear and damage
- How to fit, use, and maintain respiratory equipment
- When and how to use enclosed cabs, closed mixing systems, and other safety equipment and engineering controls
- How to secure and safely transport pesticides in a vehicle
- When medical supervision is required, and what medical supervision must be provided by the employer

Worker Training and Personal Safety

The State of California mandates that

- employee handlers receive annual pesticide safety training before they handle pesticides
- fieldworkers receive annual pesticide safety training before they enter a treated field
- pesticide safety training must be done in a language that employees understand in a location that is relatively free from distractions
- the trainer must be present for the entire training session

Below you will find the types of training required by federal and California law for workers handling or otherwise directly contacting pesticides as part of their job.

Training

California pesticide laws establish minimum standards of training for all employees working in any capacity on the farm where you work (see Appendix B for references to specific regulations). Those who handle pesticides as part of their work require additional training specific to the pesticides they will be using, which is detailed in Sidebar 7-1. This mandatory training must be delivered by a qualified trainer (listed in "Employer Responsibilities," below) who must address the following topics.

Using Pesticides Safely

- Why you must wash work clothing before wearing it again, wash the clothing separately from other laundry, and wear clean work clothing daily
- Why you must wash hands thoroughly before eating, smoking, drinking, or using the bathroom, and why you must shower thoroughly at the end of the exposure period

- Why you must never take pesticides or pesticide containers home
- What pesticide safety information is contained on Safety Data Sheets

Emergencies and Health

- Where people are likely to encounter pesticides or pesticide residues in their work environment
- How to notify workers and others so they do not enter restricted areas
- How to reduce potential hazards to children and pregnant women from pesticide exposures.
- How pesticides enter the body (skin, eyes, lungs, mouth)
- How to recognize symptoms of pesticide overexposure
- How to tell the difference among chronic, acute, delayed, and sensitization effects of pesticides
- How to administer first aid and implement emergency decontamination procedures
- Where to find the name, address, and telephone number of the clinic, physician, or hospital emergency room that can provide immediate medical treatment and when to seek emergency medical care
- Where to find your company's policy for reporting injury or illness and obtaining medical treatment
- How to recognize and avoid heat-related illness, and what to do should it occur

Legal Information and Worker Rights

- Which laws and regulations apply to you, and why it is important that you comply with them
- Why workers on your farm must be at least 18 years old to handle pesticides or perform early-entry tasks
- How employees or their designated representative(s) have the right to receive information about pesticides to which the employee may be exposed, and how employees are protected against being fired or other discrimination if they exercise these rights
- How employees can report suspected pesticide use violations to local or state agencies
- How to locate and access documents pertaining to your farm's hazard communication program, pesticide labels, Pesticide Safety Information Series sheets, Safety Data Sheets, pesticide use records, and other important documents

Employer Responsibilities

Describe the farm owner or manager's responsibility to provide PPE for mixing, loading, applying, and storing pesticides to employees.

The employer is responsible for providing the PPE and training that employees need according to pesticide labels and California regulations. These regulations cover all employees tasked with mixing, loading, applying, storing, or otherwise handling pesticides, either as a certified applicator or under the supervision of a certified applicator. The employer must make this equipment easy to access and cannot force workers to purchase it for themselves. The employer is also responsible for ensuring that PPE is cleaned and maintained, either by an employee or a third party. If you need training information, contact your local county agricultural commissioner's office.

In addition, all agricultural fieldworkers entering treated areas within 30 days of the expiration of any restricted-entry interval must receive pesticide-related training, including an explanation of pertinent state and federal laws and regulations. Qualified trainers must train fieldworkers and pesticide handlers who work in agricultural areas annually. Qualified trainers include

- pest control advisers (PCAs)
- certified private or commercial applicators
- registered foresters
- University of California farm advisors
- certain county biologists
- people who have attended a DPR-approved instructor training program

In addition, you must maintain a record of trainings delivered to fieldworkers and handlers (see Appendix C for sample recordkeeping forms).

Explain how personal protective equipment (PPE) can protect a person from hazards associated with pesticides.

List various PPE that pesticide handlers use to protect themselves from pesticide exposure.

Explain how to select the most effective PPE for the job, including understanding its limits to protect you.

Describe how to wear, clean, maintain, and store reusable PPE, and how to dispose of worn or single-use PPE.

Personal Protective Equipment

PPE offers various levels of protection, depending on the type of resistant material used. Some items of PPE simply act as barriers by keeping dry or liquid material off the skin. Others offer better protection against water-based products. Some offer protection from the chemicals that make up a concentrated pesticide product. Some types of PPE are reusable, meaning they can be used repeatedly until they become worn (though they must be cleaned at the end of each work period before you can wear them again). Other types of PPE are disposable, meaning they can only be worn once and must be discarded at the end of each work period. Disposable PPE cannot be cleaned.

WORK CLOTHES FOR HANDLERS AND FIELDWORKERS

Ordinary shirts, pants, shoes, and other work clothes are not considered PPE, even though pesticide labels often indicate that specific items of work clothing should be worn during certain activities. In California, if you handle pesticides you should wear at least a long-sleeved shirt, long pants, socks, and closed-toed shoes, even if the label does not require these items (Fig. 7-1). If you work in an area where you may contact pesticide residues, you should wear at least a long-sleeved shirt, long pants, socks, and closed-toed shoes. Make sure the long-sleeved shirt and long pants are made of sturdy material and are free of holes and tears. Fasten the shirt collar fully to protect the lower part of your neck. Fabric with a tighter weave provides better protection, but it will still absorb liquids. Nonwoven fabrics take longer to absorb liquids.

Chemical-Resistant Clothing. The term *chemical resistant* means that no measurable movement of the pesticide through the material occurs during the period of use. Some PPE is water resistant only. *Water resistant* refers to PPE that keeps a small amount of fine spray particles or small liquid splashes from penetrating the clothing and reaching the skin. *Waterproof* (liquid-proof) material keeps water-soluble materials out, but it may not necessarily keep out oil solvent–based products. Waterproof materials include items made of plastic or rubber. Read the PPE packaging carefully to determine whether the protective item is chemical resistant, liquid proof, or water resistant. Be sure linings of protective clothing are made of nonabsorbent materials to prevent pesticide contamination (Fig. 7-2).

FIGURE 7-1.

If you handle pesticides, you should wear a long-sleeved shirt, long pants, socks, and closed-toed shoes, even if the label does not require these items. California regulations require employee handlers to wear eye protection and gloves when mixing and loading, even if the label does not require these items.

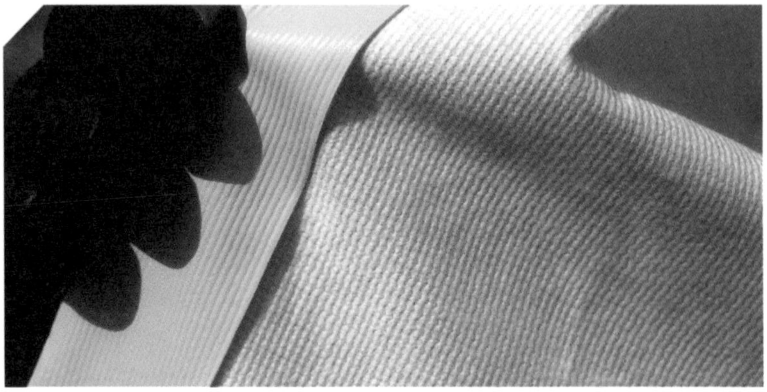

FIGURE 7-2.

Be sure linings of protective clothing are made of nonabsorbent materials to prevent pesticide contamination.

When making a decision about which PPE to use, follow these general guidelines.

- Cotton, leather, canvas, and other absorbent materials are not resistant to chemicals, even when used with dry formulations.
- Powders and dusts sometimes move through cotton and other woven materials as quickly as liquid formulations; they also may remain in the fibers even after several launderings.
- Do not use a hat that has a cloth or leather sweatband, and do not use cloth or cloth-lined gloves, footwear, or aprons.
- Cloth is very hard or impossible to clean after it becomes contaminated with pesticide, and it is usually too expensive to be disposed of and replaced after one use.

The ability of a given material to protect you from a pesticide is directly related to the type of liquid used in the formulation. Watch for signs that the material is not chemically resistant to the pesticide that you are using. Sometimes it is easy to see when plastic or rubber is not resistant to a pesticide. The material may change color, become soft or spongy, swell or bubble up, dissolve or become like jelly, crack or get holes, or become stiff or brittle. If any of these changes occur, discard the item(s) and choose another type of resistant material.

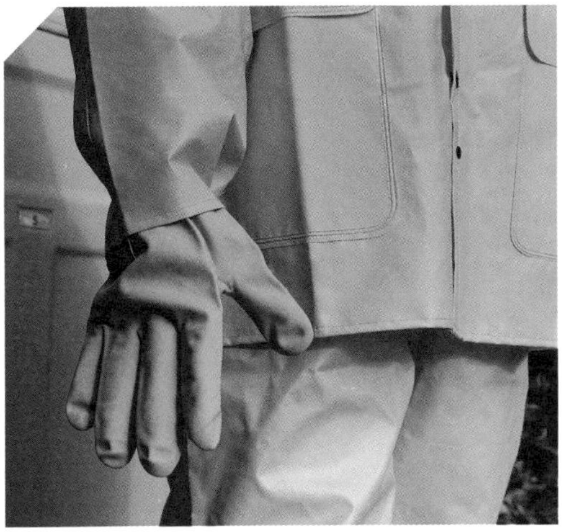

FIGURE 7-3.

Two-piece chemical-resistant coveralls must have a jacket that falls below the waist to provide adequate protection during pesticide handling activities.

COVERALLS

You should wear coveralls over a long-sleeved shirt, long pants, and socks when handling pesticides, unless otherwise specified by the label. When wearing a coverall, close the opening securely so your entire body (except the feet, hands, neck, and head) is covered. When wearing a two-piece outfit, do not tuck the shirt or coat in at the waist—have the shirt extend well below the waist of the pants and fit loosely around the hips (Fig. 7-3). Make sure coveralls are made of sturdy material such as Tyvek or laminated Tyvek unless the label requires coveralls made of some other material. Remember that the protection offered by chemical-resistant coveralls depends on the fabric and design features, such as flaps over zippers (Fig. 7-4A), elastic at the wrists and ankles (Fig. 7-4B), and seams that are bound and sealed.

A

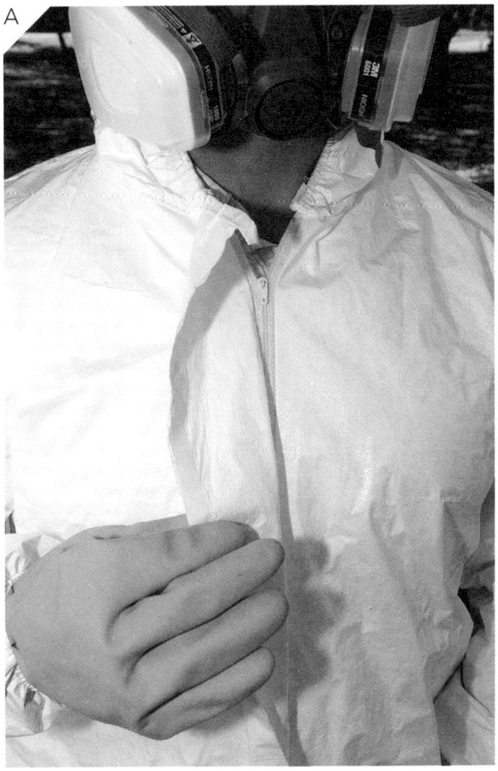

B

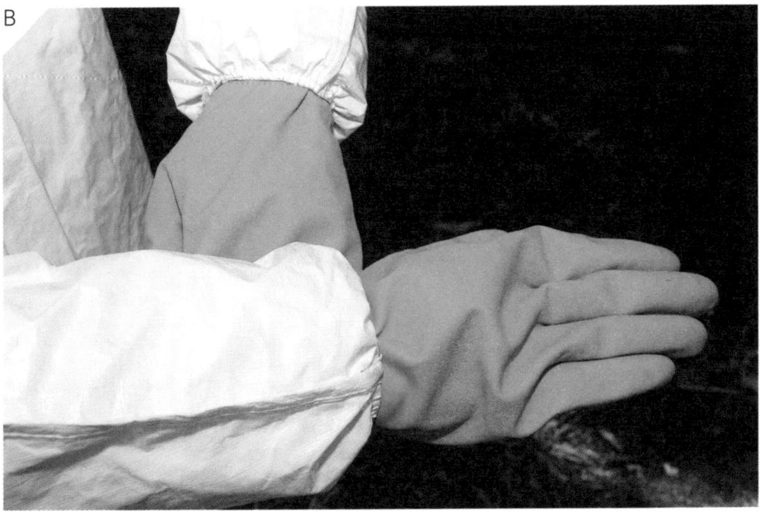

FIGURE 7-4.

The protection offered by chemical-resistant coveralls depends on the fabric and design features such as flaps over zippers (A), elastic at the wrists and ankles (B), and seams that are bound and sealed.

Several factors determine how well a coverall protects you. First, the coverall needs to fit loosely. Each layer of clothing and each layer of air between the pesticide and your skin provides added protection. That is why the coverall needs to fit loosely. If the coverall fits too tightly, there will not be a protective layer of air between it, your work clothes, and your skin.

Chemical-Resistant Suits

Some product labels require the handler to wear a chemical-resistant suit. This usually means the pesticide is very hazardous because of either acute or chronic effects. In these instances, take extra care to prevent the pesticide from getting on your skin.

The biggest drawback to chemical-resistant suits is they may make you feel hot. Unless you are handling pesticides in cool or climate-controlled environments, heat-related illness becomes a major concern. If a pesticide's label requires a chemical-resistant suit, you cannot handle the pesticide when the daytime temperature is above 80°F or the nighttime temperature is above 85°F. You are exempt from these temperature requirements if you apply the pesticide from an air-conditioned cab. Handling activities can also be performed if you wear a body-cooling device, such as an ice vest, underneath your PPE. Take extra care to avoid heat-related illness when

- temperatures go above the regulatory maximum
- temperatures reach unsafe levels and cause symptoms of heat-related illness (even if below the regulatory maximum)

Drink lots of water and take regular rest breaks to cool down when temperatures are high.

Chemical-Resistant Apron

An apron protects you from splashes, spills, and billowing dust, and it protects your coveralls or other clothing. The product label may require that you wear a chemical-resistant apron when mixing or loading a pesticide or cleaning application equipment. Even if the label does not require an apron, you should consider wearing one whenever you handle pesticide concentrates.

Choose an apron long enough to cover you from your neck to at least your knees (Fig. 7-5). Some aprons have attached sleeves and gloves. This style of apron protects your arms and hands and the front of your body by eliminating the gap where the sleeve and glove (or sleeve and apron) meet.

FIGURE 7-5.

Follow label instructions for use of waterproof or chemical-resistant aprons. Select a style with a wide bib to provide added protection.

An apron can pose a safety hazard when you are working around equipment with moving parts, such as a tractor's power take-off (PTO). If an apron can get caught in machinery, wear a chemical-resistant suit instead. Discard your apron if it develops tears or holes.

Gloves

The hands and forearms are the areas most likely to get exposed to pesticides when you are at work. Research shows that people who mix pesticides receive 85% of the total exposure to their hands and 13% to their forearms. Wearing chemical-resistant gloves, the study finds, lowers exposure by 99%. Besides the safety gains, California regulations require that employee handlers wear chemical-resistant gloves when mixing and loading any type of pesticide. Employers must provide employee handlers with chemical-resistant gloves made of the material specified on the pesticide labeling. If the label does not specify that gloves are required, or if it only states that chemical-resistant or waterproof gloves are required, employee handlers can choose to wear gloves of any chemical-resistant material, provided they are the appropriate thickness.

Gloves made from leather or fabric absorb water and pesticides, so you should only use them during pesticide handling if the pesticide label requires them. In most situations, you should choose gloves made from natural rubber, barrier laminate, butyl, nitrile, polyethylene, PVC, Viton, or neoprene (Fig. 7-6) as specified on the label. Choose a material that offers the best resistance to the pesticide you are using—thicker materials offer better protection (California regulations require most chemical-resistant gloves to be at least 14 mils). Figure 7-7 shows glove requirement statements you might find on a label. Choose materials that resist puncturing and abrasion. DPR has developed a glove key code chart printed on a wallet-sized plastic card (Fig. 7-8) to help you select the best gloves for your situation.

Wear unlined gloves, since fabrics used for built-in liners may absorb pesticides. Built-in liners make gloves dangerous to use and hard to clean. You can use woven, removable glove liners to keep your hands warm or to absorb sweat unless the label states otherwise. A glove liner must not extend beyond the end of the chemical-resistant glove. California regulations prohibit washing and reusing glove liners, so if you choose to use glove liners, you must discard them as soon as they become contaminated or at the end of each workday.

The gloves should be long enough to reach to your mid-forearm. Usually, wear the sleeves of your protective clothing on the outside of your gloves to keep out pesticides. Some special application situations, however, require holding one or both hands overhead while spraying liquids. In these cases, tuck the sleeve of the elevated arm inside the glove. Be careful when lowering your arm to prevent pesticides from entering the glove. There should not be a gap between the glove and your sleeve. Wear protective clothing with elastic at the wrist because it provides the best protection.

FIGURE 7-6.

Use only unlined, chemical-resistant gloves made of butyl, nitrile, neoprene, natural rubber, barrier laminate, polyethylene, polyvinyl chloride (PVC), or Viton.

Applicators and other handlers must wear:

- Long-sleeved shirt and long pants
- Chemical-resistant gloves made of: barrier laminate, butyl rubber ≥ 14 mils, nitrile rubber ≥ 14 mils, neoprene rubber ≥ 14 mils, polyvinyl chloride ≥ 14 mils, or Viton® ≥ 14 mils
- Shoes plus socks

FIGURE 7-7.

Glove requirement statement as seen on a pesticide label.

FIGURE 7-8.

DPR's simplified glove key code card. It fits easily into a wallet or pocket and also contains respirator restrictions. You can get one at your local county agricultural commissioner's office or order one from DPR directly. Note: Natural (code 5) refers to natural rubber. *Source:* DPR.

dpr Glove Category Selection Key

Label code	Materials Required by Law
A	1,2,3,4,5,6,7,8
B	1,2
C	1,2,3,4,7,8
D	1,2
E	1,3,4,8
F	1,2,3,8
G	1,8
H	1,8

Material Code

1 Laminate
2 Butyl
3 Nitrile
4 Neoprene
5 Natural
6 Polyethylene
7 PVC
8 Viton

All but Laminate and Polyethlene must be 14 mils or thicker.

Respirator Restrictions

N Type	NO OIL IN MIX	Dispose End of Day
R Type	OIL IN MIX	Dispose After 8 Hours Per Day
P Type	OIL IN MIX	Dispose End of Day
Organic Vapor		Dispose End of Day

Always follow label directions and permit conditions.

For more information, contact your local agricultural commissioner or the California Department of Pesticide Regulation
dpr Worker Health and Safety Branch
(916) 445-4222 http://cdpr.ca.gov

April 2020

FIGURE 7-9.

Chemical-resistant footwear protects your feet from pesticide exposure.

Footwear

Pesticide handlers often get pesticides on their feet, so you should always wear sturdy shoes and socks when you are around pesticides or pesticide residues. You should wear waterproof or chemical-resistant footwear when handling pesticide concentrates or making applications or when residues pose a hazard to your feet. You should not wear canvas or leather shoes when using pesticides because these materials absorb pesticides easily and cannot be decontaminated. Labels of some pesticides require the use of waterproof boots or boot coverings. Select protective footwear made from rubber or another chemical-resistant material. Choose the material that will best protect you from the pesticides you work with.

If a pesticide might get on your feet or lower legs, wear chemical-resistant boots that extend past your ankle and at least halfway up to your knee (commonly known as irrigator boots). You should wear waterproof boots if you will be entering or walking through treated areas when surfaces are still wet with spray. Waterproof footwear is available in conventional boot and overshoe styles. Wear the legs of your protective pants on the outside of your footwear to keep out any spray or spills (Fig. 7-9).

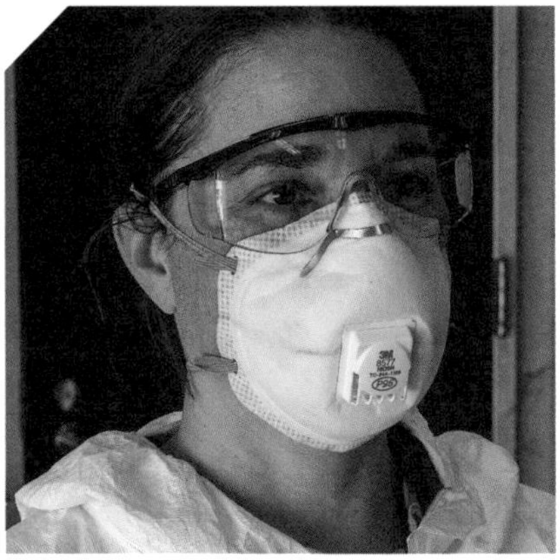

FIGURE 7-10.

Unless the label specifies the type of eyewear, you may use safety glasses that have a brow piece and side shields when handling pesticides.

Protect Your Eyes

Eyes are very sensitive to certain pesticide formulations, especially concentrates. Eyes readily absorb pesticides. In California, regulations require that employee handlers wear ANSI-approved protective eyewear during most pesticide handling activities, even if the label does not require it. Handlers usually do not need to wear eye protection when

- pesticides are being applied using a sprayer with an enclosed cab
- pesticides are being injected or incorporated into the soil
- pesticides are being applied through vehicle-mounted spray nozzles that are located below the operator with the nozzles directed downward
- vertebrate pest control baits are being applied to vertebrate burrows using equipment that keeps people from touching the material

Even in these situations, however, you must keep protective eyewear within reach in a chemical-resistant container.

Goggles, a face shield, or safety glasses with shields at both the brow and sides are examples of protective eyewear. Some labels require a particular type of eye protection. If the pesticide label does not specify the type of eye protection, you must at least wear safety glasses (marked "Z87.1" or "Z87+" in permanent, raised lettering) certified by American National Standards Institute (ANSI) that have front, side, and brow protection (Fig. 7-10, Fig. 7-11, Fig. 7-12). Avoid applying pesticides while wearing contact lenses, because contact lenses absorb pesticides and pass them through to the eye, making chronic exposure an issue.

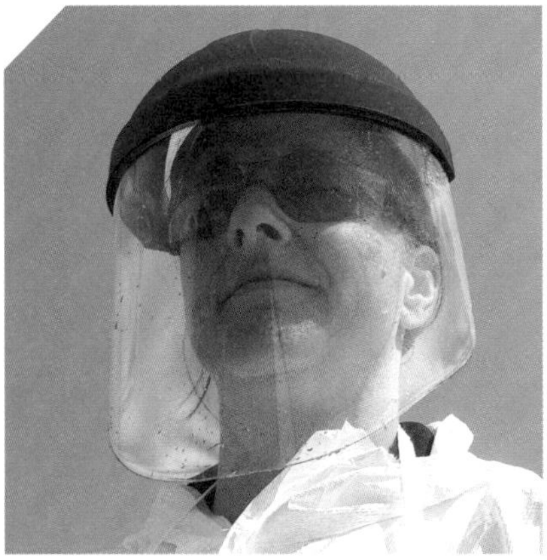

FIGURE 7-11.

Face shields provide some eye protection and keep pesticides from splashing onto your face. Wear these with safety glasses or goggles for added eye protection.

FIGURE 7-12.

Protective goggles protect the eyes during mixing and applying pesticides. Some styles allow the user to wear prescription glasses.

When the label requires protective eyewear, employee handlers must keep 1 pint of water (or eyewash) with them at all times. Water (or eyewash) can be carried on your person, or it may reside next to you in an enclosed cab or aircraft. When mixing and loading pesticides with labels that require protective eyewear or when using a closed mixing system, an eyewash station must be provided at the mix/load site. See Chapter 12 for a detailed description of what to do if pesticide gets in your eyes. Specific regulations for decontamination are listed in Appendix D.

Protect Your Respiratory Tract

The respiratory tract consists of the lungs and other parts of the breathing system. It is much more absorbent than the skin. If pesticide labels, regulations, workplace policies, or permit conditions require respirator use, employee handlers must be seen by a doctor and declared medically fit before they can wear a respirator on the job.

When employee use of respiratory protection is required by the pesticide label, restricted material use permit condition, regulation, or by workplace policy, employers must protect the health and well-being of employees by doing the following things:

- Select an individual to be the "Respirator Program Administrator" (see Glossary).
- Prepare a written respiratory protection program with work site–specific procedures for
    - selecting respirators
    - medical evaluations of employees
    - fit testing procedures for tight-fitting respirators
    - proper use in routine and emergency situations
    - cleaning, storing, inspecting, repairing, maintaining, and replacing respirators
    - air-supplying respirators (if applicable)
    - training employees on immediately dangerous to life and health (IDLH) atmospheres (if applicable)
    - training employees on the proper use of respirators
    - effectiveness evaluation
- After a written program is in place, work with a physician or other licensed health care professional to determine medical fitness of employees to wear a respirator or any conditions that the employee must follow to wear a respirator.
- Train employees on the use of the respirator, and retrain them annually thereafter.
- Fit test employees on the respirators they will be wearing prior to use and annually thereafter.
- Maintain records of the program for 3 years.
- Document annual consultations with employees on the effectiveness of the program.
- Annually review the program, making adjustments as necessary.

Respiratory Equipment

Respirators need to fit properly to be effective and safe. They should be in good working condition and be cleaned after each day's exposure period. Facial hair keeps tight-fitting respirators from protecting you, because they cannot be properly fitted to your face. Regulations prohibit pesticide applicators with facial hair from wearing tight-fitting respirators for this reason. Regulations also require that all tight-fitting respirators be fit tested to the actual wearer before use. A trained individual must perform these fit tests. Different types of respirators are listed in Table 7-1.

Cleaning and Maintaining PPE

Always keep PPE in good working condition. PPE is effective only as long as it is free from pesticide contamination and works properly. Therefore, you must frequently clean and inspect this equipment. Replace or repair equipment when you find a problem.

When you finish any work in which you are handling or are exposed to pesticides, remove your PPE right away. Start by washing the outside of your gloves with detergent and water before

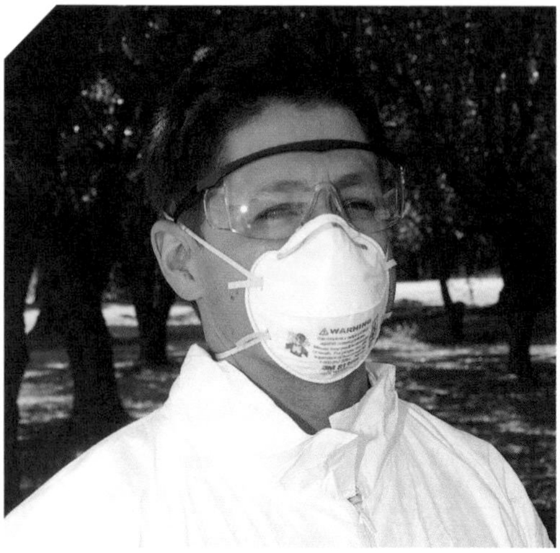

FIGURE 7-13.

Some pesticide labels require the use of NIOSH -approved dust/mist filtering respirators.

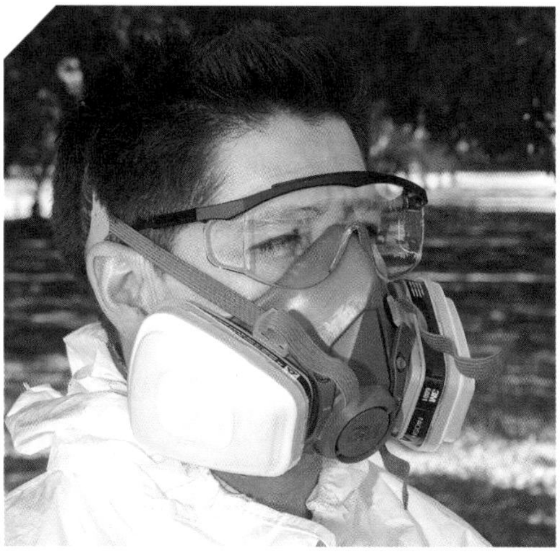

FIGURE 7-14.

For certain pesticides, labels require that you wear an organic vapor-removing cartridge respirator with a prefilter approved for pesticides. These must be NIOSH approved.

Table 7-1: Types of respirators

Respirator type*	Description
single-use filtering face pieces (dust/mist masks, particulate respirators) (Fig. 7-13)	Lightweight, soft, fairly comfortable, with two elastic straps to hold them in place. Must bear the NIOSH approval number TC-84A or indicate it has been approved according to part 84 of the Code of Federal Regulations. These types of respirators must be fit tested in the same manner as other tight-fitting respirators.
cartridge respirators (Fig. 7-14)	Fitted rubber face pieces, two-stage cartridge filters (limited effective life) specified for a particular contaminant, a one-way exhalation valve, and at least two adjustable elastic headbands. Cartridge respirators come in half-face and full-face mask styles. Used to remove low levels of pesticide vapors. Dusts and mists can also be filtered out by adding a dust/mist prefilter. Chemical cartridge respirators absorb harmful vapors or gasses and usually have an external dust/mist filter (Fig. 7-15).
canister respirators	Fitted rubber face pieces, single, often called a gas mask. Canister respirators come mainly in full-face mask styles, and they are required when using aluminum phosphide. Otherwise, the features and capabilities are similar to those of cartridge respirators.
powered air-purifying respirators (Fig. 7-16)	Forces filtered air through a hose to a hood, helmet, or face mask. The motor, pump, batteries, and filters are worn on a waist belt. Used for lengthy application jobs. People with facial hair can wear hoods, helmets, or other loose-fitting styles.
air- or atmosphere-supplying respirators (self-contained breathing apparatus, supplied-air respirator) (Fig. 7-17)	Protects you when working with concentrated amounts of highly toxic pesticides or fumigants. Provides air from either pressurized tanks (limited quantity of air) or long hoses connected to an air pump (limited range). Comes in half-face and full-face pieces, or has a hood or helmet with a clear plastic face shield. A pressure demand regulator admits fresh air on inhale for masks; hoods have a continuous flow of air around the entire head. Hoods can be worn by people with facial hair or eyeglasses. Proper cleaning and maintenance are essential to safe operation.

Note: *Eye protection must be worn with any half-face respirator. Be sure you can wear the respirator comfortably with your choice of eye protection.

A

B

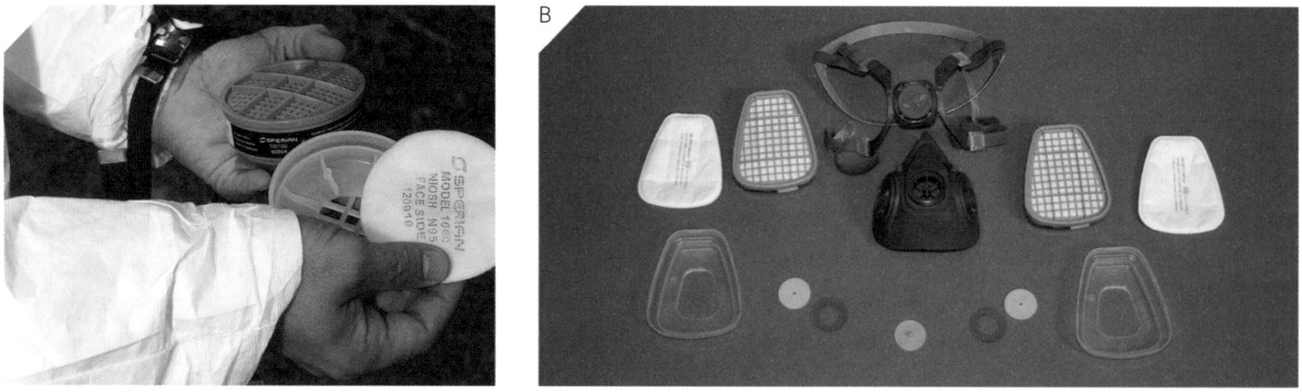

FIGURE 7-15.

Many cartridge respirators have removable filtering cartridges (A, B). Be sure the cartridges you use are approved for the type of pesticide you are applying. For example, cartridges designated "OV" can be used to protect you from most pesticides.

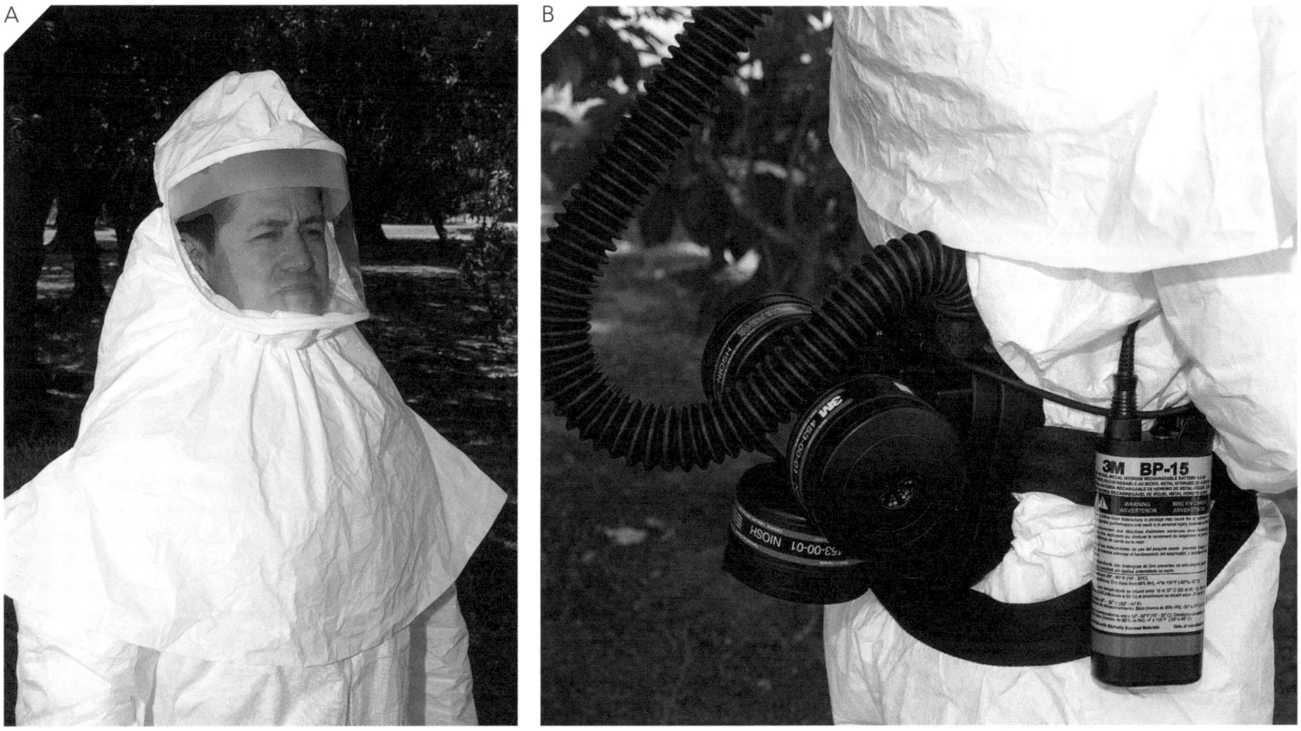

FIGURE 7-16.

A battery-powered fan forces filtered air through a flexible hose into a hood (A) in this powered air-purifying respirator. This design allows you to wear eyeglasses. People with beards and long sideburns can use this style of respirator. The filters, motor, and battery pack are worn on a waist belt (B).

FIGURE 7-17.

This self-contained breathing apparatus (SCBA) provides the wearer with uncontaminated air from a compressed air tank.

removing the rest of your PPE. Wash the outside of other chemical-resistant items before you remove your gloves. This practice helps you avoid contacting the contaminated part of the items while you are removing them. If any other clothes have pesticides on them, change them also. Determine whether the contaminated items should be disposed of or cleaned for reuse.

Chemical-Resistant Clothing

Do not rewear contaminated chemical-resistant clothing until you have washed it. Wash contaminated garments at the end of each workday. Washing right away reduces the chances of you or others being exposed to any residues. Throw away clothing that has been soaked with pesticides; send all contaminated clothing to a disposal site that can accept items that have pesticide residues on them. Clean medium to lightly contaminated clothing by washing. Throw away all single-use (disposable) chemical-resistant clothing at the end of each workday.

Change out of contaminated clothing at your work site if you can. Empty pockets and cuffs of clothing to remove any pesticide residue. Until they can be washed, place contaminated clothing into a clean plastic bag. Never reuse plastic bags, since they may acquire a buildup of pesticide residues. Do not mix contaminated clothing with any other laundry before, during, or after washing.

Work Clothing. Using hot water and liquid laundry detergent, machine wash your work clothing separately from your family's clothing. Liquid detergent removes oil-based pesticides better than powdered detergent. Sidebar 7-2A describes the best methods for washing pesticide-contaminated clothing.

SIDEBAR 7-2

Techniques for Washing Pesticide-Contaminated Clothing and PPE

A. PROCEDURE FOR WASHING PESTICIDE-CONTAMINATED CLOTHING

1. Keep pesticide-contaminated clothing separate from all other laundry.
2. Do not handle contaminated clothing with bare hands; wear rubber gloves or shake clothing from plastic bag into washer.
3. Wash only small amounts of clothing at a time. Do not combine clothing contaminated with different pesticides; wash these in separate loads.
4. Before washing, presoak clothing:
    a. Soak in tub or automatic washer or spray garments outdoors with a garden hose.
    b. Use a commercial solvent to soak product or apply prewash spray or liquid laundry detergent to soiled spots.
5. Wash garments in a washing machine using the hottest water temperature, full water level, and normal (12-minute) wash cycle. Use the maximum recommended amount of liquid laundry detergent. Never use bleach or ammonia to wash contaminated clothing—they do not remove most pesticides, and when mixed with pesticides, they release toxic vapors that can kill you.
6. If you notice pesticide odor, visible spots, or stains, repeat step 5 several times until clothing is fully clean.
7. Clean washing machine before using for other laundry by repeating step 5, using full amount of hot water, normal wash cycle, and laundry detergent, but no clothing.
8. Hang laundry outdoors on a clothesline to avoid contaminating automatic dryer.

Do not attempt to wash heavily contaminated clothing; destroy it by transporting it to an approved disposal site. These suggestions will help you reduce contamination of your family's laundry with pesticides:

1. Whenever possible, wear disposable protective clothing that can be destroyed after use.
2. Always wear all required protective clothing when working with pesticides.
3. Wear clean protective clothing daily when working with pesticides. Wash contaminated clothing daily.
4. Remove contaminated clothing at work site and empty pockets and cuffs. Place clothing in a clean plastic bag until it can be laundered. Keep contaminated clothing separated from all other laundry.
5. Remove clothing immediately if it has had a pesticide concentrate spilled on it.

B. PROCEDURE FOR WASHING CONTAMINATED PPE

1. Wash only a few items at a time so there is plenty of agitation and water for dilution.
2. Wash items separately from all other laundry in a washing machine, using heavy-duty liquid detergent and hot water, on the longest heavy-duty cycle. Use two rinse cycles.
3. Use two entire machine cycles to wash items that are moderately to heavily contaminated. (If PPE is too contaminated, bundle it in a plastic bag, label the bag, and take it to a household hazardous waste collection site.)
4. Run the washer through at least one additional entire cycle without clothing, using detergent and hot water to clean the machine before any other laundry is washed.

Boots and Gloves

Wash rubber boots and gloves under running water to remove pesticide residues before you take them off. Use detergent mixed with water and a soft brush, then rinse with clean water (Fig. 7-18). Do not get the insides of the boots wet. At the end of each day, wash rubber gloves with soap and warm water. Inspect them for holes while washing and discard the gloves if you find any. You may wash gloves (separated from other laundry) in a washing machine by placing them into a cloth net bag. Use warm water and wash according to the instructions given below for protective clothing. Turn gloves inside out for drying. Store dry boots and gloves in plastic bags to keep them clean and prevent deterioration. Remember that glove liners cannot be washed and reused–they must be thrown away at the end of each pesticide handling activity.

FIGURE 7-18.

Wash boots before removing them. Use a brush and soapy water, then rinse with clean water. Do not get the boots wet inside. Let boots air out after washing; store them in a clean plastic bag once they dry.

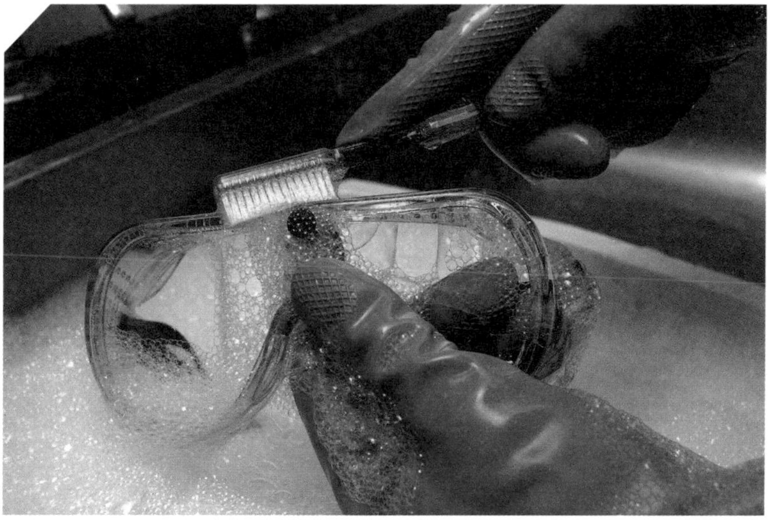

FIGURE 7-19.

Wash safety glasses, goggles, and face shields in warm, soapy water. Use a soft brush or cloth to remove pesticide residue. Blot dry and store in a clean plastic bag.

Face Shields and Goggles

Wash goggles, face shields, shielded safety glasses, and respirator bodies and face pieces after each day of use. Use care when washing face shields and goggles to avoid scratching the plastic. Submerge them in warm, soapy water. You can remove pesticide residue using a soft, wet cloth or soft brush (Fig. 7-19). Do not rub antifogging lenses, since this reduces their effectiveness. Rinse well with clear water and air-dry or blot with a soft cotton cloth; rubbing may cause scratching. Inspect goggles and face shields for scratches, cracks, and loss of headband elasticity. You can replace scratched lenses on many styles rather than buying new goggles. Store goggles and face shields in paper bags to keep them clean.

Respirators

Inspection. You should inspect your respirator for wear and damage twice: once before using it for the first time each day and again before cleaning it at the end of each day. Check headbands for fraying, tears, or loss of elasticity, and replace them if necessary. Remove filters and replace the gaskets if they are brittle, broken, or warped. Valve assemblies are essential parts of a cartridge respirator and must be in good working order. Take apart and inspect valve flaps for wear, deformities, or punctures. Replace parts if you think they might leak. Check the threads of all valves and cartridge parts to make sure they are in good condition and that the valve seats are smooth. Look for cracks and scratches.

Examine the face piece for cracks, cuts, scratches, and any signs of aging. If you find damage, replace the parts.

When replacing items on a respirator, use only approved replacement parts for that brand and model. If you use unapproved parts, the respirator is not in compliance with the law and may put you at risk. If you keep a respirator for emergency use or as a backup, inspect it at least monthly.

Cleaning. After removing filters from reusable cartridge respirators, soak the face piece, gaskets, and valve parts in warm water and mild liquid detergent. Do not use abrasives or cleaning compounds containing alcohol or other similar chemicals. You must use sanitizers if more than one person wears the same respirator. Use a soft brush or cloth to remove any pesticide residue (Fig. 7-20). Rinse the respirator and valve parts in clean water. Air-dry your filters rather than leaving them in direct sunlight or using applied heat. Heat and sunlight can damage filters or cause them to wear out more quickly. Removable respirator cartridges must be disposed of at the end of the work period, even if the respirator is used for an hour or two.

After it is completely dry, put the respirator back together and store it in a clean plastic bag to protect it from dirt and exposure to particles in the air.

FIGURE 7-20.

After use, remove cartridges and wash respirators in warm, soapy water. Use a soft brush or cloth to remove pesticide residue.

Dispose of one-time-use cartridge respirators according to the manufacturer's instructions. Do not try to clean disposable cartridge respirators.

If you remove your respirator between handling activities, follow these guidelines:

- Wipe the respirator body and face piece with a clean cloth.
- Replace caps, if available, over cartridges, canisters, and prefilters.
- Seal the respirator in a sturdy, airtight container, such as a plastic bag with a zip closure. If you do not seal the respirator immediately after each use, the disposable parts will have to be replaced more often. This is because cartridges and canisters continue to collect impurities as long as they are exposed to the air.

Washing PPE

Be sure that the people who clean and maintain your PPE know that touching these pesticide-contaminated items can harm them. Instruct them to wear gloves and an apron and to work in a well-ventilated area, if possible, and avoid inhaling steam from the washer or dryer.

Follow the manufacturer's instructions for cleaning chemical-resistant items. If the manufacturer directs you to clean the item but gives no detailed instructions, follow the steps in "Procedure for Washing Contaminated PPE" in Sidebar 7-2B. Some chemical-resistant items that are not flat, such as gloves, footwear, and coveralls, must be washed twice—once to clean the outside of the item and a second time after turning the item inside out. Some chemical-resistant items, such as heavy-duty boots and rigid hats or helmets, can be washed by hand using hot water and a heavy-duty liquid detergent.

Storing PPE

Never use your PPE for any purpose other than pesticide handling. Store your PPE in a clean, dry place, protected from temperature extremes and bright light when you are not using it. If possible, put the items in sealable plastic bags. Light, heat, dirt, and air pollution contribute to the breakdown of rubber, plastic, and synthetic rubber products. Never store PPE in areas where you keep pesticides.

Limits to Protection

The amount of protection provided by PPE is limited: it never completely protects you. You still must keep pesticides from being spilled, splashed, or sprayed onto your body. Your equipment helps to reduce exposure, but you must do everything possible to keep the exposure from happening.

Pesticides trapped next to your skin cannot dissipate through air movement or volatilization. Therefore, if you get pesticides on your skin or clothing before putting on PPE, the equipment may increase the amount of pesticide absorbed. You will also contaminate the inside of the protective garment. Always wear clean PPE over clean clothing.

Loose-fitting PPE can also get caught in any equipment that has moving parts, such as the PTO of a pesticide sprayer, and cause injury. Be aware of your surroundings when wearing loose-fitting PPE to prevent possible injury.

Engineering Controls

List the different kinds of engineering controls and explain when these are used.

Engineering controls protect people as they are mixing, loading, and applying pesticides. These protective devices are considered PPE, even though they are not worn on the body. Employers must provide them to employee handlers if the PPE is required by pesticide labels or California regulation. Engineering controls include enclosed cabs, closed mixing systems, pesticide packaging, and atmosphere-monitoring devices. It is important to remember that whenever you perform activities outside of these engineering controls, you must wear all required PPE. All required PPE needs to be available in a chemical-resistant container, and you must put it on as soon as you stop using the protective engineering control (as when you exit the enclosed cab of a sprayer during or immediately after an application).

Enclosed Cabs. Enclosed cabs on tractors protect you from exposure to pesticides (Fig. 7-21) while you are making an application.Some types protect you from spray droplets and mists and offer a comfortable, air-conditioned environment as well. These cabs, however, do not replace the label respirator requirements unless the specified respirator is a filtering face piece. Half- or full-face respirators of all types, if required, must be worn while making an application from inside this type of cab.

FIGURE 7-21.

Enclosed cabs protect operators against pesticide exposure. This model includes a pesticide air filtering system that eliminates the need for the operator to wear a respirator while inside the cab.

Closed Mixing Systems. Employees must use a closed mixing system when mixing, loading, diluting, or transferring liquid formulations of pesticides with certain precautionary statements on the label, as illustrated in Figure 7-22A and B. There are two types of closed mixing systems, Tier 1 and Tier 2.

- A Tier 1 system removes the pesticide from the container and rinses and drains the empty container while it is still connected to the closed system. If you mix liquid pesticides with statements on the label that say "Fatal if absorbed through the skin" or something similar, you must use a Tier 1 system.
- A Tier 2 system removes the pesticide from the container but does not rinse it. You must use a Tier 2 closed mixing system if you mix liquid pesticides with statements on the label that say "May be fatal if absorbed through the skin," "Corrosive, causes skin damage," or something similar.

Closed mixing systems enable accurate and safe measuring of pesticides being put into the spray tank. Not all situations require closed mixing systems. To find out if your situation requires the use of a closed mixing system,

A

B

PRECAUTIONARY STATEMENTS

HAZARDS TO HUMANS AND DOMESTIC ANIMALS

DANGER: Corrosive. Causes skin damage. May be fatal if absorbed through the skin. Do not get on skin or clothing. Prolonged or frequent repeated skin contact may cause allergic reactions in some individuals. Harmful if swallowed or inhaled. Irritating to eyes, nose and throat. Avoid breathing vapor or spray mist. Do not get in eyes.

FIGURE 7-22.

Closed mixing systems are required when mixing more than 1 gallon per day of liquid DANGER pesticides for the production of an agricultural commodity. These systems allow you to accurately measure the liquid pesticides. Most also rinse the empty containers (A). This label shows two of the three precautionary statements that require the use of a closed mixing system in California (B). The third one is "fatal if absorbed through the skin."

review regulations on the DPR website, cdpr.ca.gov/docs/legbills/calcode/chapter_.htm. See the DPR Pesticide Safety Information Series A-3 for the requirements for closed mixing systems.

Packaging. Special pesticide packaging helps to reduce exposure to concentrated pesticide active ingredients. This packaging includes preweighed water-soluble bags and packets of powdered formulations. These dissolve in the spray tank, reducing your exposure to the powder and dust. Pesticides packaged in this way are considered closed mixing systems according to regulations.

Chapter 7 Review Questions

1. PPE protects you from exposure to pesticides by ____________.

☐ a. keeping dry and liquid materials off your skin
☐ b. covering only the most vulnerable part of your body
☐ c. preventing you from having workplace accidents

2. Pesticide handlers must be trained in which three subject areas?

☐ a. integrated pest management, pest identification, and application equipment maintenance
☐ b. closed mixing systems, PPE requirements, and reading the pesticide label
☐ c. using pesticides safely, emergencies and health, and legal information and worker rights

3. Who is responsible for purchasing, cleaning, and maintaining PPE required by pesticide labels?

☐ a. workers
☐ b. employers
☐ c. manufacturers

4. Match the PPE with the protection it offers.

1. coverall	a. worn directly over your work clothes (long-sleeved shirt, long pants, and socks) to protect them from pesticide contamination
2. chemical-resistant suit	b. protects your coveralls, and guards you from splashes, spills, and billowing dust
3. chemical-resistant apron	c. protects your lungs from pesticides in the air
4. chemical-resistant hat	d. protects you when a large amount of pesticide could be deposited on your clothing over an extended period of time
5. gloves	e. protects your eyes, and prevents liquids from splashing onto your face during mixing
6. face shield	f. protects you from overhead exposure or exposure to a lot of airborne particles
7. respirator	g. keeps pesticides from contaminating your hands and forearms

5. Match the situation with the most appropriate PPE.

1. You are spraying a large volume of a DANGER pesticide that is likely to drift onto your clothing and may remain in the air as you make the application. Temperatures are moderate.	a. chemical-resistant headgear, goggles, gloves, and coveralls
2. You are spraying a CAUTION pesticide over your head into trees.	b. closed mixing system, coveralls, safety glasses, respiratory protection, and gloves
3. You are mixing and then loading a dust pesticide with a precautionary statement saying "fatal if absorbed through skin."	c. chemical-resistant suit, respiratory protection, goggles, gloves, and a hat

6. True or false?

☐ True ☐ False a. Reusable PPE must be cleaned at the end of each work period, before using the equipment again.

☐ True ☐ False b. An individual fit test is required to ensure that your respirator fits properly and works effectively to protect you.

☐ True ☐ False c. Nonwoven coveralls and hoods marked "disposable" can be worn for as many as 7 workdays.

☐ True ☐ False d. Avoid heat-related illness by using less PPE than required and making the application as quickly as possible.

☐ True ☐ False e. PPE can worsen an exposure incident if you put it on over clothing that had been contaminated by pesticides and not properly cleaned.

7. Engineering controls that help protect you from pesticide exposure include which of the following?

☐ a. enclosed cabs and closed mixing systems

☐ b. SCBA devices and water-soluble pesticide packaging

☐ c. chemical-resistant materials and atmosphere-monitoring devices

Chapter 8
Using Pesticides Safely

Pesticide Applicator Safety 96
Safe Application Methods 104
Cleaning Application Equipment 106
Personal Cleanup 106
Chapter 8 Review Questions 107

Knowledge Expectations

1. Describe ways in which applicators ensure the public's safety before, during, and after pesticide applications.
2. Explain why and in which situations it is important to communicate with people in the area before making a pesticide application.
3. Describe how to restrict access to areas where pesticides are in use or have been used.
4. List procedures and safety precautions for transporting pesticides in a vehicle.
5. List the components of a proper storage area.
6. Describe methods for mixing and loading pesticides safely, including the equipment, location, and procedures used in the process.
7. Explain how to properly process all types of pesticide containers for disposal.
8. Describe how to identify potentially sensitive areas that could be adversely affected by pesticide application, mixing and loading, storage, disposal, and equipment washing.
9. Describe the procedures to follow for safe, effective cleanup after handling pesticides, including cleaning application equipment, as well as personal decontamination.

Pesticide Applicator Safety

You are the key to preventing pesticide accidents. By following the pesticide label and following pesticide laws and regulations, you can avoid most problems. In addition, when mixing pesticides together or with other materials, test in a small jar to be sure that the combination will be safe. Also, check your equipment to be sure it is working well. Remember, faulty, broken, or worn equipment causes accidents. Finally, never take alcohol or drugs before, during, or immediately after applying pesticides.

This chapter describes how to prevent exposure to pesticides while you work. You avoid pesticide-related problems by

- reading and following the directions in the pesticide labeling
- complying with all laws, regulations, and restricted material permit conditions that apply to pesticide handling, storage, and application in your work situation
- using safe work habits
- wearing the required personal protective equipment (PPE)
- protecting people from pesticide exposure
- avoiding practices that may harm nontarget plants and animals
- keeping pesticides on target

Planning Applications to Ensure Safety

Describe ways in which applicators ensure the public's safety before, during, and after pesticide applications.

Planning for pesticide applications helps to prevent accidents. First, find out about the pesticides you use by studying Safety Data Sheets and pesticide labels. From these you will learn about the dangers of exposure and the actions you can take to avoid it. Inspect areas where you will be working to find potential hazards that can affect your safety. Finally, plan what you will do if an accident happens. Use the checklist in Sidebar 8-1 to help you plan future pesticide applications.

Planning for Accidents

Plan for the possibility of an accident. This process includes finding an appropriate medical facility before you need emergency care. Also, find out where to get help with spill cleanup. Post in your vehicle or prominently at the work site the name, address, and telephone number of the medical facility closest to where you are working. Also, write down the telephone numbers of the local fire department, sheriff, and highway patrol (see Sidebar 12-1 in Chapter 12) and keep a copy of the label, which you can bring with you if necessary. The emergency number 9-1-1 usually gives you immediate access to medical help, local fire services, and law enforcement agencies; however, it is not a replacement for the phone numbers you should have and provide to employee handlers and others at the work site.

Plan what to do if there is a pesticide spill, and be ready to protect the public from danger. Know the proper first aid to give to victims of pesticide exposure or heat-related illness. Understand the steps you must take to reduce injury to yourself and others in case of an accident. Be sure you have enough emergency water for washing your eyes and skin. For more about planning for and responding to emergencies, see Chapter 12.

Explain why and in which situations it is important to communicate with people in the area before making a pesticide application.

Describe how to restrict access to areas where pesticides are in use or have been used.

Protecting People at or near the Application Site

In agriculture, fieldworkers may be working near where you are making a pesticide application. You must protect these fieldworkers from any type of pesticide exposure. Do not allow workers into an area that is being treated with a pesticide. Employers also must keep fieldworkers from entering the application exclusion zones (AEZs) created by that application. In addition, workers are not allowed into treated fields during restricted-entry intervals unless they are trained as early-entry workers (workers who enter an area after the pesticide application is complete, but before the restricted-entry interval or other entry restriction has ended). If your application may impact others in the area (residents, etc.), you should make sure they know to stay indoors during

SIDEBAR 8-1

Checklist for Planning a Pesticide Application

PERSONAL

- ☐ Medical checkup and necessary blood tests?
- ☐ Properly trained for this type of application?

PESTICIDE

- ☐ Read and thoroughly understood label?
- ☐ Checked to be sure use is consistent with target pest and application area?
- ☐ Read Safety Data Sheet for information on hazards?
- ☐ Obtained necessary certifications and permits?
- ☐ Know proper rate of pesticide to be applied?

EQUIPMENT

- ☐ Proper personal protective equipment (boots, gloves, respiratory equipment, protective clothing, eye protection, head wear)?
- ☐ Necessary measuring and mixing equipment?
- ☐ Suitable application equipment for this job (tank capacity, pressure range, volume of output, nozzle size, pump compatible with formulation type)?
- ☐ Application equipment properly calibrated?
- ☐ Emergency water and first aid supplies?
- ☐ Necessary supplies to contain spills or leaks (absorbent materials, cleaning supplies, holding containers)?

TRANSPORTING

- ☐ Safe transport of pesticides to application site?
- ☐ Pesticides and containers secured from theft or unauthorized access?
- ☐ Vehicles properly marked and permits obtained, if necessary, for transporting hazardous materials and hazardous wastes?

MIXING AND LOADING

- ☐ Safe mixing and loading site located?
- ☐ Clean water available for mixing?
- ☐ Water pH tested?
- ☐ Proper adjuvants obtained for correcting pH, preventing foaming, and improving deposition?
- ☐ Compatibility of pesticide tank mixes or fertilizer-pesticide combinations checked?
- ☐ Liquid containers triple-rinsed and rinsate put into spray tank?

TREATMENT SITE

- ☐ Boundaries of treatment site inspected?
- ☐ Environmentally sensitive areas within and around treatment area identified?
- ☐ In agricultural applications, people working or living in or near treatment area, including fieldworkers and their supervisors, notified?
- ☐ Treatment site properly posted with required signs?
- ☐ Soil types determined and noted, if these are factors in pesticide efficacy?
- ☐ Livestock, pets, honey bees, other animals properly protected?
- ☐ Location and depth of groundwater determined, if applicable?
- ☐ Hazards within treatment site identified, including electrical wires and outlets, ignition sources, obstacles, steep slopes, and other dangerous conditions?
- ☐ Plants in treatment area in proper condition for pesticide application (correct growth stage, not under moisture stress, other requirements as specified on pesticide label)?

WEATHER CONDITIONS

- ☐ Weather suitable for application (low wind, proper temperature, lack of fog or rainfall)?

APPLICATION

- ☐ Established application pattern suitable for treatment area, hazards, and prevailing weather conditions?
- ☐ Application rate will give most uniform coverage?
- ☐ Equipment frequently checked during application to ensure that everything is working properly and providing a uniform application?

CLEANUP

- ☐ Application equipment properly cleaned and decontaminated after application?
- ☐ Personal protective equipment safely stored and cleaned or laundered according to approved methods?
- ☐ Disposable materials disposed of in approved way?

DISPOSAL

- ☐ Paper pesticide containers disposed of according to local regulation?
- ☐ Plastic and metal containers triple-rinsed?
- ☐ Plastic and metal containers properly stored until properly disposed of or recycled?

STORAGE

- ☐ Storage facility suitable for pesticides?
- ☐ Unused pesticides returned to supplier or stored in locked facility for later use?

REPORTS

- ☐ Necessary reports filed with requesting agency?

FOLLOW-UP

- ☐ Treatment areas inspected after application to ensure that pesticide controlled the target pests without causing undue damage to nontarget organisms or surfaces in treatment area?

DAMAGE

- ☐ Damage, if found, promptly reported?

the application period. Make pesticide applications at times when workers and others are not present nearby. These times may include very early morning, late afternoon, or during the night. Preventing drift also lessens exposure risks to fieldworkers in adjacent areas.

Restricted-Entry Interval

A restricted-entry interval (REI) is the time that must go by after a pesticide application before anyone (other than trained early-entry workers) can enter the treated area. Pesticide labels and the California Code of Regulations Title 3 (3 CCR) list REIs. Always review the pesticide labeling as well as 3 CCR for the REIs that apply in your situation. Pesticide use recommendations written by licensed pest control advisers must indicate the required REI. The local agricultural commissioner can also give you this information. When the REI on the label differs from California requirements, the longest REI applies. Also, when tank-mixing pesticides with different REIs, you must use the longest interval listed. You should check with the county agricultural commissioner's office for special REIs when you apply mixtures of certain pesticides. Even if the label does not require the posting of warning signs, you must post if the REI exceeds 48 hours.

Posting

Sometimes you must post treated areas with warning signs (Fig. 8-1). Posting is a way to notify employees and others about a treated area and its buffer zones. Regulations require posting signs to be made of a durable material. Signs should be printed in English and a non-English language that is understood by a majority of workers. Signs must contain a skull and crossbones at the center and the word DANGER in letters big enough to read from a distance of 25 feet. If the restricted-entry interval is more than 7 days, the sign must also list

- the name of the pesticide
- the date the restricted-entry interval ends
- the property operator's name
- any field identification

FIGURE 8-1.

Some pesticide labels require treated areas to be posted. Also, an area must be posted if the restricted-entry interval is more than 48 hours. All greenhouse applications must be posted when the restricted-entry interval is greater than 4 hours, unless access is carefully controlled throughout the restricted-entry interval.

If you are applying Category I pesticides through an irrigation system (chemigation), you must post an additional warning sign. Check pesticide labels and current federal, state, and local laws to determine requirements for posting. Local offices of county agricultural commissioners have this information.

To post a treated area, place signs at usual points of entry. If there are no clear points of entry, signs must be posted on corners of the treated area. Along unfenced areas next to roads and other public rights-of-way, signs should be no more than 600 feet apart. Post the area before you make an application (but no sooner than 24 hours before the application). Signs must remain in place throughout the REI. Remove them within 3 days after the end of the REI and before you allow workers to enter the treated area. See Table 8-1 for a list of signs, the information they must contain, and the situations in which each must be posted.

Notification

At agricultural sites, before applying any pesticide you must notify all employees of the farming operation who are working within ¼ mile of the treatment area. Do this orally or by posting warning signs unless the pesticide label specifies the method you must use. As a reminder, you

TABLE 8-1:

Warning signs in English and Spanish and when to post them

Signs	Information required	When to post
	the words "danger pesticides" a skull and crossbones pictogram the words "keep out"	if you will apply pesticide in a greenhouse or other entirely enclosed space if you will apply pesticide in an enclosed space with an REI of more than 4 hours if the label-required REI is more than 48 hours if the label requires field posting
	the words "danger pesticides" a skull and crossbones pictogram the words "keep out" the name of the pesticide the name of the grower the REI expiration date	if the REI is more than 7 days
 and 	the words "danger pesticides" a skull and crossbones pictogram the words "keep out" the name of the pesticide the name of the grower the REI expiration date and the words "keep out" a stop sign pictogram with the word "stop" the words "pesticides in irrigation water"	if the REI is more than 7 days and pesticides are applied in irrigation water (chemigation)
 and 	the words "danger pesticides" a skull and crossbones pictogram the words "keep out" and the words "keep out" a stop sign pictogram with the word "stop" the words "pesticides in irrigation water"	if you will apply pesticide in a greenhouse or other entirely enclosed space through an irrigation system (chemigation) if you will apply pesticide in an enclosed space through an irrigation system (chemigation) and the REI is more than 4 hours if the pesticide is applied in irrigation water (chemigation) and the label requires field posting or the REI for the chemigated pesticide is more than 48 hours
	the word "danger" the words "poison storage area" the words "all unauthorized persons keep out" the words "keep door locked when not in use"	if you store DANGER or WARNING pesticides

must post warning signs if the REI exceeds 48 hours, even if the label doesn't specify the need for posting. Tell workers when you plan to make the application so they will leave and then not reenter the treated area. Tell them what pesticides you will apply and describe the hazards if they should become exposed. Also tell them when they can reenter the area.

Transporting Pesticides

List procedures and safety precautions for transporting pesticides in a vehicle.

Vehicles used to transport pesticides should be in good mechanical condition, including power train, chassis, and any onboard bulk tanks and fittings. In particular, make sure safety and control components such as brakes, tires, and steering are in good working order. A poorly maintained vehicle is, by itself, a safety risk; adding pesticides to the picture increases the potential risk of injury or contamination should an accident happen. Always carry equipment needed to make repairs in case there is a problem while the vehicle is in use.

Never carry pesticides in the passenger compartment of a vehicle because spilled chemicals and hazardous fumes can seriously injure the occupants. Spilled pesticides can be hard or impossible to remove from the vehicle's interior, leading to long-term exposures. If pesticides must be carried in a station wagon, utility van, or similarly enclosed vehicle, ventilate the cargo and passenger compartments and keep passengers and pets away from pesticides during transport. Remember, cargo can shift during collisions and other sudden stops. Placing a safety barrier between passenger and cargo areas is a good idea.

FIGURE 8-2.

Transport pesticides in the cargo area of your vehicle, never in the passenger area. Secure containers in the cargo area and protect them from moisture and damage. Never carry people, animals, food, animal feed, or clothing in the same area.

The cargo area must be able to securely hold containers and protect them from tears and punctures or crashes that could damage them (Fig. 8-2). Enclosed cargo boxes provide the greatest protection but are not always practical. Cargo boxes also offer the added benefit of security from curious children, thieves, or vandals. Open truck beds are convenient for loading and unloading, but take precautions to minimize the possibility of theft or losing containers on sharp turns or bumpy roads. Never stack pesticide containers higher than the sides of the vehicle. Make sure flatbed trucks have side and tail racks and tie-down rings, cleats, or racks to simplify the job of securing the load. Before loading, inspect the cargo area for nails, stones, or sharp edges or objects that could damage containers. Steel beds are better than wood because they are more easily cleaned after a spill. Devices are available for some vehicles that protect pesticide cargo in the case of a rear-end collision. In California, you are required to secure pesticides in a way that will prevent spillage onto or off the vehicle. Paper, cardboard, and similar containers must be covered (when necessary) to keep them from getting wet. If you are transporting a service container, make sure its label includes the name and address of the person responsible for the container, the name of the pesticide, and the pesticide's signal word, according to regulations.

Vehicle Operator

The person who is driving the transport vehicle can be held accountable for any injuries, contamination, or damage caused by a pesticide spill. The driver may be the only person able to react to a spill. In some instances, the driver may need to assist first-response emergency personnel as they arrive on the scene. In California, the driver is responsible for contacting emergency response agencies and for ensuring the cleanup of spilled material if there is an accident. Before transporting any pesticide product, determine

- who you should contact in an emergency
- what to report to governmental agencies and emergency response personnel
- how to clean up (or who to contact to clean up) a pesticide spill

Everyone involved in transporting pesticides should receive training in basic emergency response procedures, including spill control and emergency notification procedures. Refer to Chapter 12 for specific information on how to respond to a fire, spill, or leak involving pesticides.

Before leaving the farm with a pesticide cargo, make sure that the technical data for all pesticide products and emergency information for spill response are in the vehicle. For information on transporting pesticides on public highways in California and to get regulatory updates, contact the following agencies:

- California Highway Patrol
- California Public Utilities Commission
- The Governor's Office of Emergency Services

Sidebar 8-2 lists places to get information on and check regulations covering the transportation of pesticides.

Storing Pesticides

List the components of a proper storage area.

Although many farmers use existing buildings or areas within existing buildings for pesticide storage, it is always best to build a separate storage facility just for pesticides. You should store pesticides in enclosed areas, on an impermeable (concrete) surface, and protected from rain. A well-designed and -maintained pesticide storage site

- protects people and animals from exposure
- reduces the chance of environmental contamination
- prevents damage to pesticides from temperature extremes and excess moisture
- safeguards the pesticides from theft, vandalism, and unauthorized use
- reduces the likelihood of liability

SIDEBAR 8-2

Where to Get Information and Regulations on Transporting Pesticides

FOR INTERSTATE MOVEMENT OF PESTICIDES

U.S. Department of Transportation

California Field Office
1325 J Street, Suite 1540
Sacramento, CA 95814-2941
(916) 930-2760

FOR TRANSPORTATION OF PESTICIDES WITHIN CALIFORNIA

California Highway Patrol Motor Carrier Safety Unit
Division Offices

Northern Division
2485 Sonoma Street
Redding, CA 96001-3026
(530) 225-2715

Valley Division
2555 First Avenue
Sacramento, CA 95818
(916) 731-6300

Golden Gate Division
1551 Benicia Road
Vallejo, CA 94591-7568
(707) 551-4180

Central Division
5179 North Gates Avenue
Fresno, CA 93722-6414
(559) 277-7250

Southern Division
411 N. Central Avenue, Suite 410
Glendale, CA 91203
(818) 240-8200

Border Division
9330 Farnham Street
San Diego, CA 92123-1216
(858) 650-3600

Coastal Division
4115 Broad Street, #B-10
San Luis Obispo, CA 93401-7963
(805) 549-3261

Inland Division
847 E. Brier Drive
San Bernardino, CA 92408-2820
(909) 806-2400

A proper pesticide storage area will require that you do all of the following:

- Post warning signs visible from at least 25 feet away on doors and windows to alert people if there are DANGER or WARNING pesticides stored inside (see Table 8-1).
- Keep points of entry securely locked so pesticides cannot be stolen or accessed by unauthorized people.
- Make sure that runoff will not be a problem and that the area is not prone to flooding (i.e., not near a stream or other body of water).
- Check to make sure that the area is at least 100 feet from a well to prevent accidental groundwater contamination.
- Make sure that the area is well ventilated and insulated or temperature controlled to regulate air quality and temperature.
- Keep pesticides out of direct sunlight to prevent overheating.
- Use spark-proof lighting fixtures and switches to provide good lighting for pesticide handlers working in the storage area.
- Choose flooring and shelving that is made from impermeable materials that are free of cracks and easy to clean after a spill or leak.
- Use secondary containment bins to limit leaks and spills to a confined space.
- Store only pesticide containers, equipment, and a spill kit at the site. Keep food and PPE elsewhere.
- Keep pesticide containers with their labels in plain sight, and make sure all labels are legible.

Mixing Pesticides Safely

Describe methods for mixing and loading pesticides safely, including the equipment, location, and procedures used in the process.

Methods for mixing pesticides safely are the same for large and small volumes. You must

- read the mixing directions on labels of all pesticides you will be using
- determine what PPE you need for mixing and application, and ask for what you don't already have on hand
- check the spray equipment for cracked hoses or leaks, and make sure the filters, screens, and nozzles are clean
- have enough fresh water nearby for eye flushing and to supply decontamination facilities for washing your entire body in case of an accident
- choose the proper order to add chemicals, including adjuvants, to the spray tank (Chapter 11 describes correct mixing order)

Closed Mixing and Loading Systems

In California, you are required to use a Tier 1 or Tier 2 closed mixing system when employees mix or load liquid pesticides with high acute dermal toxicity, which is indicated by any of the following precautionary statements on the label:

- Fatal if absorbed through skin (Tier 1).
- May be fatal if absorbed through skin (Tier 2).
- Corrosive, causes skin damage (Tier 2).

There are two main types of closed mixing and loading systems. One type uses mechanical devices to deliver the pesticide from the container to the equipment. The other type uses water-soluble packaging. For more about water-soluble packaging, see "Water-Soluble Bags or Packets (WSB or WSP)" in Table 3-5. For more about Tier 1 and 2 closed mixing systems, including PPE requirements, see "Engineering Controls" in Chapter 7.

Mixing and Loading Pesticides

Pesticide packages are available in different units of weight or volume. Whenever you can, plan a mixture that uses an even, preweighed amount of pesticide. The unit cost may be higher when you buy pesticides in smaller packages. However, the cost is small when the convenience and added safety of not having to weigh or measure is taken into account. Do not open pesticides packaged in water-soluble packets, since these may contain highly hazardous formulations. Calibrate application equipment to use whole packets.

Select a mixing site that is at least 100 feet from any unprotected wellheads and that you can clean easily in case of an accident. When not using premeasured packets, measure and weigh pesticides in a clear, open area. If outdoors, stand upwind to reduce chances of exposure. Read the pesticide label for specific PPE required for mixing and loading pesticides. No matter what the label says, you must wear a face shield, goggles, or other protective eyewear and chemical-resistant gloves when performing mixing and loading activities. Always measure and pour pesticides below eye level to reduce chances of spills or splashes into your face and eyes (Fig. 8-3).

After measuring or weighing the correct amount of pesticide, carefully pour it into the partially filled (no more than three-quarters full) spray tank (Fig. 8-4). Rinse the measuring container and pour the rinse solution into the spray tank. Use caution while rinsing to prevent splashing. Many closed mixing systems have container-rinsing devices that pump the rinse solution into the pesticide tank. Unless rinsed automatically, drain liquid containers into the spray tank for 30 seconds after you empty them. Rinse and drain the containers three more times (triple-rinse). Sidebar 8-3 has easy-to-follow instructions for proper triple-rinsing of pesticide containers. As an added precaution, you must puncture all triple-rinsed containers to prevent reuse. Even after triple-rinsing, you cannot leave these containers unsecured, so store them in your secure pesticide storage area until you can bring them to a pesticide container recycling center or a site that accepts nonhazardous wastes (such as a Class II landfill).

Explain how to properly process all types of pesticide containers for disposal.

For bags that hold dry pesticides, follow these emptying guidelines:

- Open and empty the bag so that no pesticide material remains in the bag that can be poured, drained, or otherwise removed.
- Empty the pesticide bag completely, and hold the bag upside down for 5 seconds after continuous flow stops.
- Straighten out the seams so that the bag is in its original "flat" position.
- Shake the bag twice and hold it upside down for 5 seconds after continuous flow stops.
- Check with the county agricultural commissioner to find out how to properly dispose of pesticide bags.

FIGURE 8-3.

Always pour and measure pesticides below eye level. If measuring outdoors, stand upwind. Wear the label-mandated PPE for mixers when measuring pesticides.

FIGURE 8-4.

Carefully pour pesticides into the spray tank. Rinse measuring containers and empty and triple-rinse liquid pesticide containers. Pour the rinse solutions into the spray tank.

SIDEBAR 8-3

TRIPLE-RINSING PROCEDURES FOR PESTICIDE CONTAINERS

PROCEDURE

1. When container is empty, let it drain into spray or mixing tank for at least 30 seconds.
2. Add the correct amount of water to container as follows:

Container size	Rinse solution needed
5 gallons or less	¼ of the container volume
more than 5 gallons	⅕ of the container volume
28 gallons or more	does not require triple-rinsing—return to dealer

3. Close container.
4. Shake container or roll to get solution on all interior surfaces.
5. Drain container into sprayer or mixing tank. After empty, let it drain for an additional 30 seconds.
6. Perform steps 2 through 5 two more times.
7. Puncture container to prevent reuse.

AMOUNT OF ACTIVE INGREDIENT REMOVED FROM A 5-GALLON CONTAINER BY TRIPLE-RINSING

Rinse step	Amount of active ingredient remaining*
drain	14.1875 grams a.i.
1st rinse	0.2183 gram a.i.
2nd rinse	0.0034 gram a.i.
3rd rinse	0.00005 gram a.i.

Note: *After draining, a 5-gallon container is assumed to still contain 1 ounce of formulated pesticide. This would amount to 14.1875 grams of a.i. if the formulation contained 4 pounds of a.i. per gallon.

FIGURE 8-5.

When filling a spray tank, be sure there is an air gap between the filler pipe and the top level of the water in the tank. This prevents backflow of pesticide-contaminated water into the water supply.

Do not allow the spray tank to overflow during filling. Also, never let the hose, pipe, or other filling device touch the liquid in the tank. If you fill the tank through a top opening, leave an air gap between the spray tank and filling device for backflow prevention (Fig. 8-5). This space should equal at least twice the diameter of the filling pipe. Using an air gap stops the spray mixture from siphoning back into the water supply after you stop the water flow. Side- or bottom-filling systems require check valves to prevent backflow of pesticides into the water supply. For effective mixing methods, see "Mixing Pesticides" in Chapter 11.

SAFE APPLICATION METHODS

To use pesticides safely as well as effectively, keep them within the treatment area and apply them in the right amounts. Avoid spills, leaks, and drift, which waste the pesticide and may leave residues in nontarget areas. Calibrate application equipment regularly, since poor equipment calibration can result in too little or too much pesticide reaching the target site. Safe pesticide applications require that you

- use proper equipment
- develop good application methods
- reduce or eliminate drift
- be aware of all potential hazards

Safe application methods require that you

- work with the weather
- control droplet size and deposition
- know the application site and its hazards
- develop application patterns for the site that avoid hazards and hazardous environmental conditions
- leave buffer zones (strips) to protect sensitive areas

Describe how to identify potentially sensitive areas that could be adversely affected by pesticide application, mixing and loading, storage, disposal, and equipment washing.

Working with the Weather

Weather can significantly influence the safety and effectiveness of pesticide applications in agricultural areas. Its effect on pesticide applications in greenhouses and other confined spaces is more subtle. Table 8-2 explains various weather conditions and associated hazards.

Site Characteristics and Environmental Hazards

Carefully observe the site where a pesticide will be applied, noting the environmentally sensitive areas where pesticide drift, leaching, runoff, and residues will do the most damage. Sensitive areas include ponds, streams, or marshes and areas where water moves easily, such as watersheds. These

TABLE 8-2:

Various weather conditions and their associated hazards

Weather conditions	Description	Associated hazards
air temperature	Temperatures above 80–85 degrees are considered high. The pesticide label will tell you if a product can be applied at high temperatures. Temperatures below 40 degrees are considered low. The pesticide label will tell you if a product may become unstable at low temperatures.	high temperatures: • pesticide breakdown (degradation) • pesticide volatilization • phytotoxicity (plant damage) high to mild temperatures: • damage to honey bees and other pollinators that are active within this temperature range low temperatures (below 40 degrees): • chemical destabilization that reduces effectiveness
temperature inversion	Warm air 20–100 feet or more above the ground forms a cap that blocks vertical air movement (Fig. 8-6). Fine spray droplets and pesticide vapors get trapped in the inversion layer, where they become concentrated.	long-distance drift of fine pesticide droplets or vapors
precipitation, fog, heavy dew	Precipitation comes in various forms, such as rain, sleet, and snow. Clouds of tiny water droplets at or near the earth's surface are called fog. When water vapor has condensed into droplets (usually overnight), it forms dew.	dilution and degradation of pesticides • material washed off treated surfaces • runoff into surface water • leaching into groundwater • drift in fog clouds
sunny and clear	Ultraviolet light is produced by the sun and is most intense during sunny, clear weather.	pesticide breakdown (degradation)
wind speed	Wind speed is the measured velocity of air moving in the atmosphere.	wind speeds above 10 miles per hour: • drift • volatilization • uneven pesticide deposition wind speeds under 3 miles per hour: • uneven pesticide deposition • possible inversion condition

FIGURE 8-6.

A temperature inversion is caused by a layer of warm air occurring above cooler air close to the ground. This warm air prevents air near the ground from rising, similar to a lid.

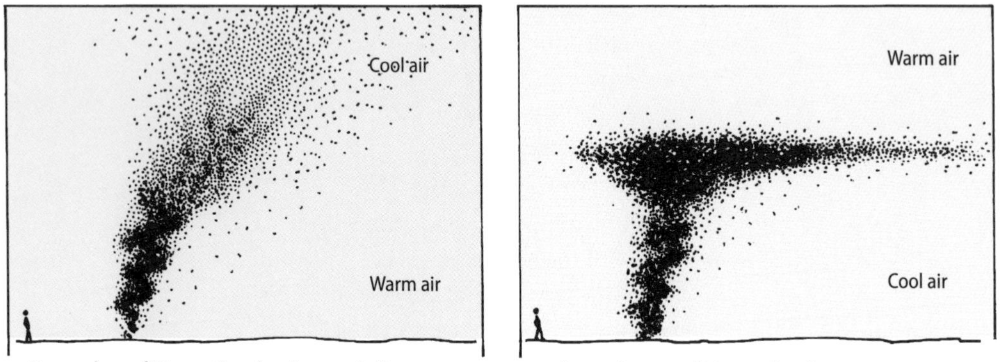

are easy to see when observing an area targeted for pesticide application and should be avoided. Pesticide labels often indicate a minimum distance from surface water (a buffer zone) for safe application. Sensitive areas also include fields above aquifers or near sinkholes or wells. Since some sensitive areas, such as aquifers, cannot be seen when observing an application site, it is best to check DPR's Ground Water Protection Area (GWPA) maps (see the California Department of Pesticide Regulation website, cdpr.ca.gov/docs/emon/grndwtr/gwpamaps.htm) to find out whether leaching, runoff, or both are major problems. Always check with your local county agricultural commissioner if the property you're treating is near or in a GWPA to ensure that it is actually within the GWPA. Whenever groundwater is present, however, be sure to check the pesticide label and California's Groundwater Protection List to see whether the product is likely to leach. If so, you should not apply that product to that site.

Sensitive areas also include homes, schools, hospitals, parks, playgrounds, commercial areas, and other places where people may work, live, or play. The location and type of crops, especially those next to the application site, must also be considered when you are selecting pesticides. If neighboring crops are not listed on the pesticide label and the product drifts onto and contaminates them, those crops will have to be destroyed.

Cleaning Application Equipment

Describe the procedures to follow for safe, effective cleanup after handling pesticides, including cleaning application equipment, as well as personal decontamination.

After each use, you must clean and decontaminate application equipment. Otherwise, residues remaining in tanks may contaminate the next pesticide mixture. Pesticide residue on the outside of application equipment can be hazardous to people who must operate or repair it. Therefore, wash the outside of spray equipment with water, using a small amount of detergent if needed. Clean equipment either in an area where you can contain runoff or at the application site.

Personal Cleanup

After using pesticides, clean your PPE, shower thoroughly, and change into clean, uncontaminated clothing. When showering, take special care to wash your hair and clean your fingernails. Place clothing that you wore during the pesticide application into a plastic bag until you can wash it. Never eat, drink, smoke, or use the bathroom until you have thoroughly washed. Always wash your work clothing separately from other clothing, especially if you bring it home to wash. You can find additional information on cleaning and maintaining work clothing and PPE in Chapter 7.

Chapter 8 Review Questions

1. **Which situation requires you to notify people in the area of your pesticide application?**
   - ☐ a. when employees are working within ¼ mile of the treatment site
   - ☐ b. when employees hang laundry to dry outside of nearby homes
   - ☐ c. when employees' families have gardens that could be affected by drift

2. **The restricted-entry interval (REI) for the pesticide you are applying is 8 days. How will you keep people out of the treated field during that period?**
   - ☐ a. Place warning signs at usual points of entry, or in the case of an unfenced field, at the corners of the treated area.
   - ☐ b. Notify fieldworkers of the REI orally before the application and remind them again after the application.
   - ☐ c. Erect a temporary barrier around the treated area that remains locked for the duration of the REI.

3. **When transporting pesticides in a vehicle, you should __________.**
   - ☐ a. secure the packages inside the passenger compartment
   - ☐ b. carry them in the cargo area of a truck, but have someone ride in that area to make sure containers remain undamaged in transit
   - ☐ c. secure containers in the vehicle's cargo area after checking it carefully for anything that might damage containers in transit

4. **A proper pesticide storage facility should be ____________.**
   - ☐ a. protected by a security system and equipped with a telephone for emergencies
   - ☐ b. securely locked and clearly identified as a pesticide storage facility
   - ☐ c. well lighted and supplied with plenty of sturdy wooden storage shelves

5. **Match the sensitive area with the steps used to protect it.**

1. lakes, ponds, or streams	a. Choose pesticides that are less likely to drift and are less toxic to people and animals. Leave a buffer strip adjoining these features of the landscape.
2. aquifers, sinkholes, or wells	b. Check the label for the proper distance to maintain from these features of the landscape, and use a formulation less likely to have problems with runoff.
3. parks, schools, playgrounds, or recreational areas	c. Check DPR's Ground Water Protection Area (GWPA) maps, and avoid using pesticides that leach to protect these features of the landscape. Read the label to find out about a pesticide's leaching potential.

6. **Which statement about triple-rinsing pesticide containers is true?**
   - ☐ a. You must wear extra PPE for triple-rinsing with a closed system.
   - ☐ b. Triple-rinsed containers can be taken to a Class II landfill.
   - ☐ c. Triple-rinsing is not necessary if you intend to recycle the container.

7. **High temperatures during or soon after a pesticide application can cause increased ____________.**
   - ☐ a. leaching potential and long-distance drift
   - ☐ b. absorption and translocation
   - ☐ c. phytotoxicity and breakdown

CHAPTER 9
APPLICATION EQUIPMENT

Application Equipment 110
Application Methods and Equipment 117
Maintaining Application Equipment 117
Chapter 9 Review Questions 125

Knowledge Expectations

1. List components of liquid application equipment, and explain how they work together.
2. Identify the components of liquid application equipment that work best with different types of pesticide formulations.
3. Describe how to recognize wear in various components.
4. Describe the various nozzles available, including design, size, angles, and output.
5. List the important factors to consider when selecting nozzles for a given application.
6. List the types of application equipment, and describe the advantages and limitations of each type.
7. List the types of application equipment used to apply liquids, and describe the situations in which each should be used.
8. List the types of application equipment used to apply dusts, and describe the situations in which each should be used.
9. List the types of application equipment used to apply granules, and describe the situations in which each should be used.
10. List types of bait application equipment, and explain how they work.
11. Describe how to maintain and clean different kinds of equipment.
12. Describe the situations in which chemigation systems can be used.

Application Equipment

List components of liquid application equipment, and explain how they work together.

Selecting, using, maintaining, and properly calibrating application equipment helps to ensure that pesticides are applied safely, accurately, and effectively. In this chapter, you will learn about the application methods and equipment used most often in agricultural settings, including an in-depth treatment of nozzles and their selection. You will also read about precision farming tools–electronic systems that help you target pesticides far more precisely than you can through traditional means.

Liquid Application Equipment

Most liquid application equipment uses hydraulic pressure or air to generate pesticide droplets and move them to the target. This equipment is either hand-operated or powered by mechanical sources such as a tractor power take-off (PTO) or an electric, gasoline, or diesel engine. Equipment used for applying nonfumigant pesticides through irrigation systems (chemigation) is also considered liquid application equipment. Chemigation equipment includes specialized components that are treated more fully in *The Safe and Effective Use of Pesticides*.

Liquid application equipment consists of several parts, including

- a tank for mixing and holding the pesticide
- a pump or other device for creating pressure to move the liquid through the equipment
- one or more nozzles for breaking the spray up into small droplets and directing it toward the target and regulating flow
- filter screens or strainers and a pressure gauge
- on some equipment: fans, pressure regulators, strainers, control valves, agitators, booms or hand spray guns, hoses, couplings and fittings, unloaders, surge chambers, chemical injectors, spray shields, and closed-system mixing equipment (Fig. 9-1).

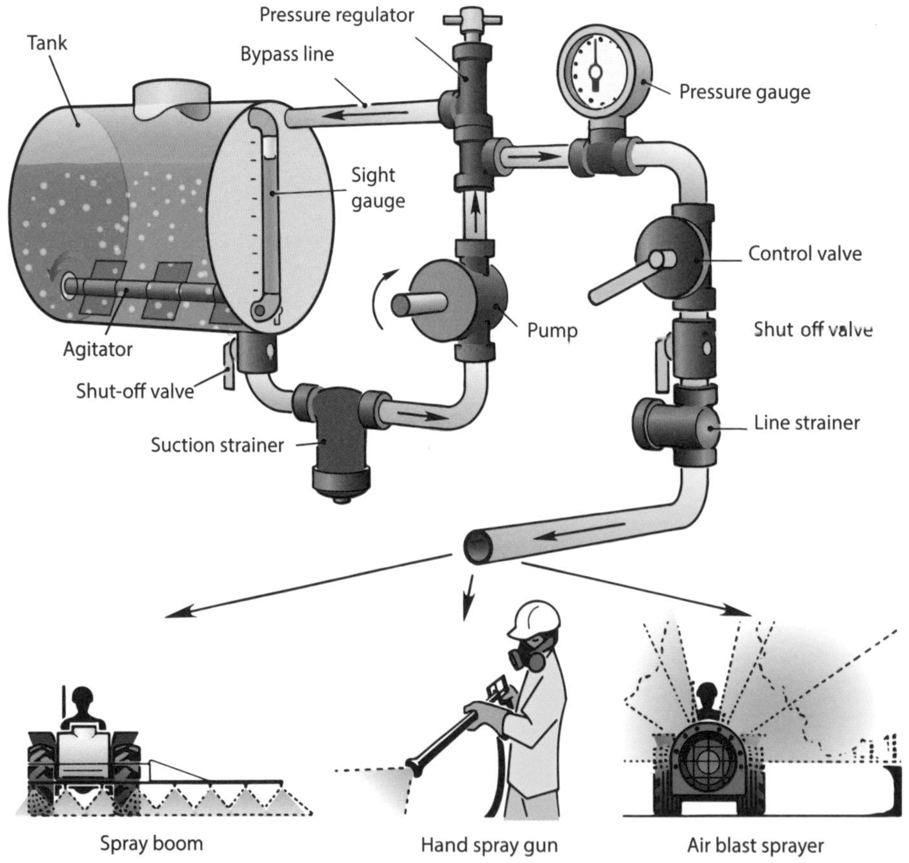

FIGURE 9-1. Liquid application equipment usually includes a tank for mixing and holding pesticides (often equipped with an agitator) and a pump for creating hydraulic pressure, and may also include a pressure regulator, pressure gauge, control valve, and several types of strainers. Spray is emitted through nozzles on a spray boom, manifold, or hand spray gun, and may be dispersed by a fan.

Any of the above components can be switched out or adjusted to accommodate changing conditions, different pests, and various formulations (Fig. 9-2), and they should be replaced immediately when they become worn or damaged.

> **Use centrifugal pumps which provide propeller shear action for dispersing and mixing this product. The pump should provide a minimum of 10 gallons per minute per 100 gallon tank size circulated through a correctly positioned sparger tube or jets.**

FIGURE 9-2.

Some labels will indicate specialized equipment and components that are needed to properly apply the material.

Equipment Components and Recognizing Wear

Identify the components of liquid application equipment that work best with different types of pesticide formulations.

Describe how to recognize wear in various components.

Tanks. Stainless steel tanks resist rusting and corrosion, and you can use them with most pesticides without problems, although they are more expensive than other types of tanks. These tanks also come in versions that are galvanized or coated, so choose carefully the type of metal tank you select, since some coatings should not be used with specific pesticide active ingredients. All metal tanks must be checked regularly to ensure they remain free from rust or corrosion, and coated tanks should also be checked to ensure there are not scratches or chips exposing bare metal to corrosive materials. Stainless steel tanks can be repaired by a skilled technician.

Fiberglass tanks are strong and durable, and are easily repaired with resin if damaged areas are small. Fiberglass tanks must be checked carefully for scratches or abrasions, since these can result in absorption of pesticide liquids. This absorption can result in contamination of future tank mixes.

Thermoplastic tanks are lightweight and durable but become flexible and will deform if they get warm. Unlike fiberglass tanks, minor scratches or abrasions do not cause absorption problems, but they are harder to repair if they become punctured or cracked. Thermoplastic tanks can also become degraded by long-term exposure to sunlight and weather. If you see spiderwebbing (small, interconnected cracks) on any surface of a thermoplastic tank, it means that the tank is beginning to weaken and should be replaced.

Pumps. Diaphragm pumps are a popular style used on several types of spray equipment. Diaphragm pumps handle abrasive and corrosive chemicals well because only the chemical-resistant diaphragm contacts pumped liquids. They are also simple to maintain and repair. Diaphragm pumps have only a few moving parts (Fig. 9-3). Diaphragms usually wear out after a while, so you must replace them when they begin to leak. You can tell there is a leak when the oil in the reservoir on the pump turns milky. The petroleum-based solvents in emulsifiable

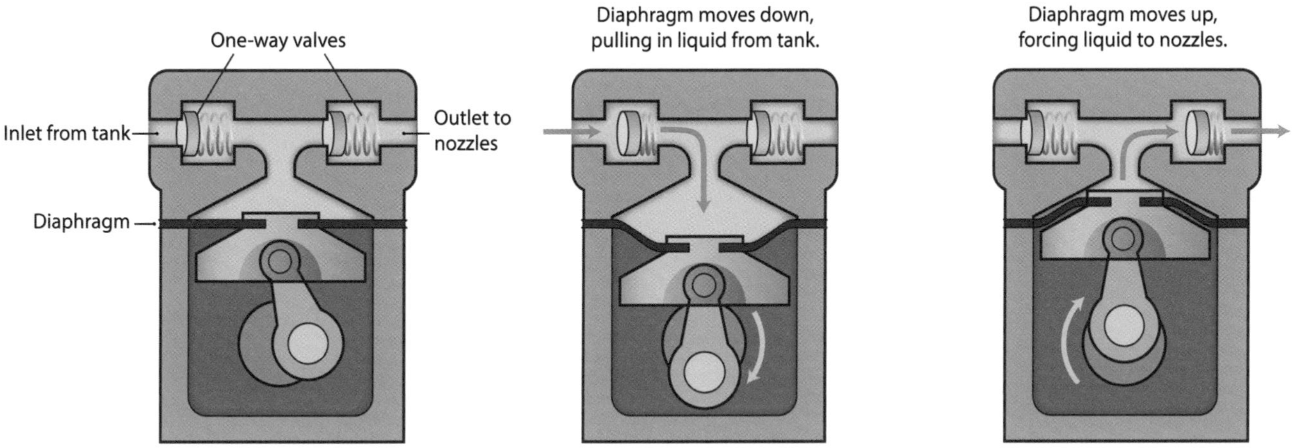

FIGURE 9-3.

In a diaphragm pump, a flexible diaphragm is moved up and down by a cam mechanism. This oscillation moves liquid through one-way valves. Some diaphragm pumps incorporate two or three diaphragms moved by the same cam.

concentrate formulations accelerate the breakdown of these rubber components. They can also tear under too much pressure. Replace the rubber valves when they fail to seal properly.

Roller pumps are among the least expensive pump types. In these pumps, a series of rollers fit into slots around the circumference of a rotating disc, or impeller (Fig. 9-4). Roller pumps are subject to considerable wear, especially from abrasive materials like wettable powders. Rollers made of rubber last longer. However, you must use nylon or Teflon rollers to pump petroleum-based pesticides such as oils or emulsions because petroleum-based pesticides deteriorate rubber. Usually, you can easily replace worn-out rollers.

Piston pumps are generally the most expensive type of pump, but they are necessary if you make a lot of high-pressure applications or use both high and low pressures using a sprayer equipped with a rate controller. Piston pumps work by forcing fluids through one-way valves as its pistons move within their cylinders (Fig. 9-5). Pulsating pressure may be a problem with piston pumps, as it is in diaphragm pumps. A surge tank can be used in these cases to even out the pulses. Abrasive chemicals cause wear in piston pumps, although most have easily replaceable cylinder liners and piston cups. More expensive piston pumps have stainless steel or ceramic cylinder liners to resist wear.

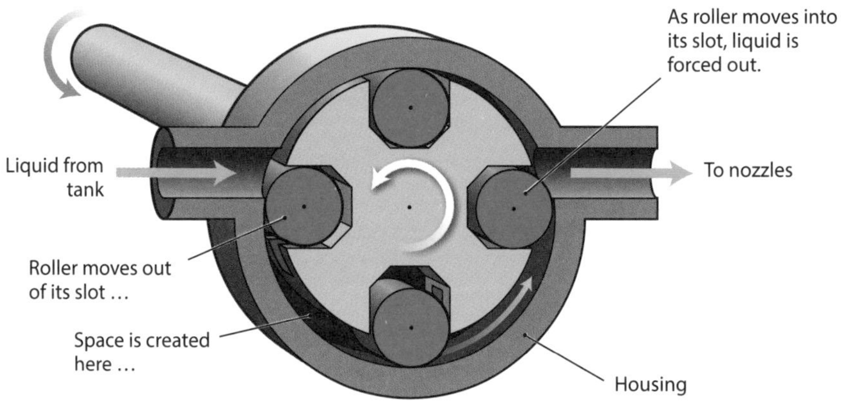

FIGURE 9-4.

Roller pumps consist of cylindrical rollers that move in or out of slots in a spinning rotor. This action creates space for liquid during half of the rotor rotation and discharges the liquid out of the pumping chamber during the remainder of the rotor rotation.

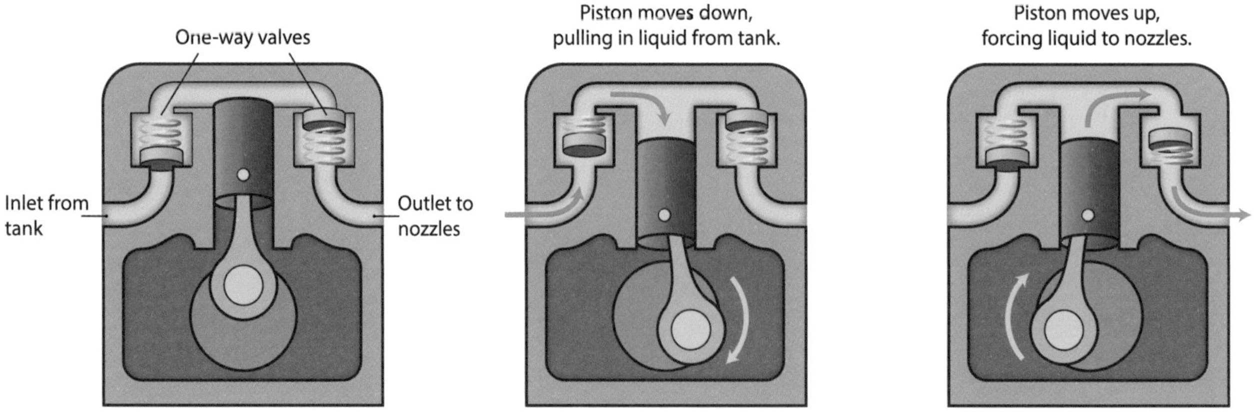

FIGURE 9-5.

This sequence shows how a piston pump works. The downward movement of a piston draws liquid through a one-way valve into the cylinder. When the piston moves up, liquid is forced out through another one-way valve. Some pumps consist of several pistons working opposite each other.

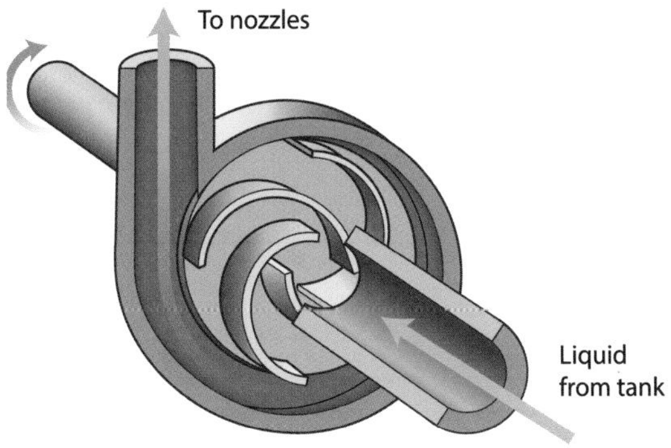

FIGURE 9-6.

In a centrifugal pump, liquid enters near the center of a vaned rotor. As the rotor spins, the liquid is moved away from the center by centrifugal force. Rotors must turn at a high rpm in order to build up sufficient pressure for most spray applications.

Manufacturers produce centrifugal pumps out of high-impact plastic, aluminum, cast iron, or bronze. These pumps are heavy-duty and adaptable to a wide variety of spray applications. You can use them for spraying abrasive materials because there is no close contact between moving parts. A high-speed impeller creates the pumping action that forces liquids out of the pump (Fig. 9-6). They are often easy to repair and work well for high-volume air blast sprayers.

Mechanical Agitators. These agitators require some maintenance, especially where shafts pass through tank walls. Packings and grease fittings prevent leaks but need periodic tightening and servicing. Be sure to use a marine-grade grease on bearings and seals exposed to liquids. Also, periodically tighten and service belts or chains.

Strainers and Filter Screens. The filter screens in strainers need to be checked to ensure they are not clogged. If you notice pressure dropping within the system and unusual strain on the pump, check your filter screens to see if they have become clogged. Typically, the strainers are located before and sometimes also after the pump, and before the nozzles. For best results, check the nozzle manufacturer's catalog for the proper nozzle screen size for the nozzle you choose.

Describe the various nozzles available, including design, size, angles, and output.

Nozzles. Manufacturers make nozzles out of various materials, all of which are subject to wear. Worn nozzles do not generate proper droplet patterns or regulate flow to the manufacturer's specifications, resulting in poor pesticide coverage and unpredictable droplet sizes (Fig. 9-7). The design of the nozzle (including the metals or plastics it is made of), the kinds of materials being sprayed, and the spray pressure all influence the amount of wear on the nozzle. Flat-fan nozzle styles with sharp-edged orifices initially wear much faster than, for example, a flooding tip with a circular orifice. Also, as the spray pattern angle increases, the wear on the

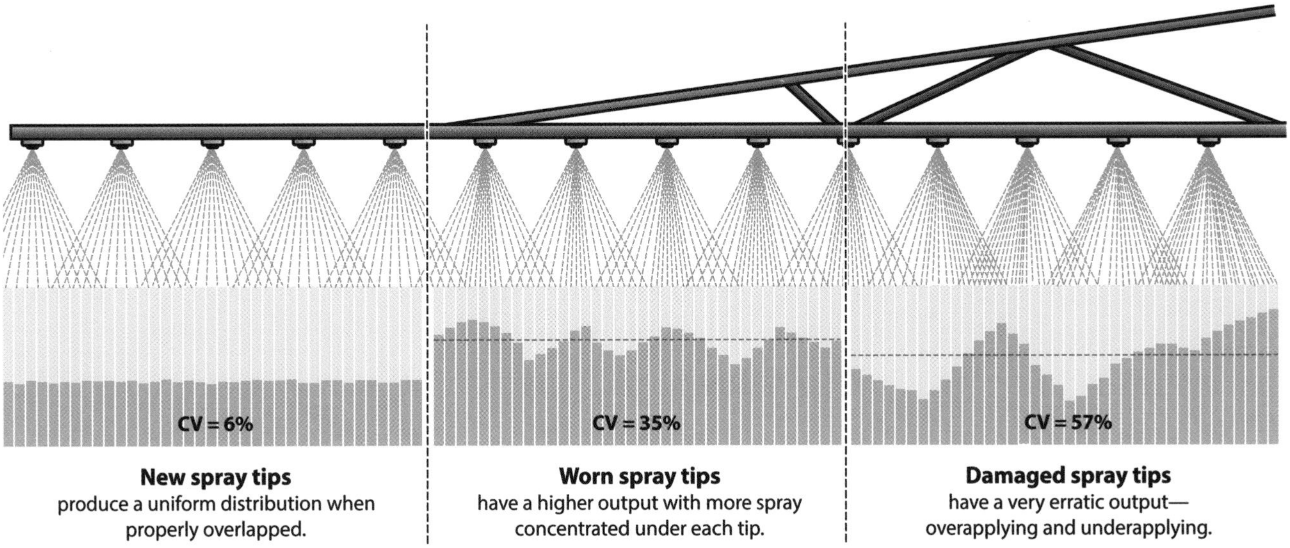

FIGURE 9-7.

Spraying with worn or damaged nozzles causes unacceptable variations in the amount of pesticide deposited in an area. This graphic illustrates what can happen when output differs by more than 10% between nozzles on the same boom (CV = coefficient of variation). *Source:* TeeJet.

nozzle increases. Further, the size of the orifice affects wear: larger orifices wear more slowly than smaller ones.

Spray materials influence wear differently depending on the amount of dissolved or suspended solids in the liquid. True solutions (like mixtures made using soluble powder formulations) cause the least amount of wear, while suspensions (like mixtures made using wettable powders) cause more wear.

The solids that influence wear may be any ingredient in the formulation. Rate of nozzle wear, even when using the same type of pesticide over time, varies. Sometimes chemical companies make small changes in their formulations that have no effect on the performance of the pesticide but may influence nozzle wear. Also, formulations of the same pesticide can vary from one manufacturer to another. Some pesticides form crystals under certain conditions of water pH, water temperature, and the presence of other chemicals. These crystals often increase wear on nozzles. Higher liquid pressure increases the rate of nozzle wear, as well.

As a nozzle wears, the volume and pattern of spray change and affect the application. Replace nozzles when they fail to deliver either accurate pesticide amounts or the desired spray pattern.

The output from nozzles of the same size, used together on a boom, should not vary from each other by more than 10%. If they do, it is an indication that your nozzles should be replaced. To ensure uniform wear, be sure to use nozzles made from the same material. Nozzles may be made from any of the following materials.

- Brass nozzles are moderately inexpensive but wear quickly from abrasion. Brass is an acceptable material if you do not use abrasive sprays or if you replace nozzles frequently.
- Stainless steel nozzles do not corrode, and they resist abrasion. Although hardened stainless steel wears exceptionally well, these nozzles are more expensive than most others. To address the issue of expense, some manufacturers produce plastic nozzles with stainless steel inserts, reducing the cost while increasing the life of the nozzle.
- Aluminum and monel (nickel alloy) nozzles resist corrosion but are highly susceptible to abrasion because they are made of such soft metals. Avoid using aluminum and monel nozzles unless you need specific corrosion resistance.
- Plastic nozzles are the least expensive. The plastic material resists corrosion, but nozzles made totally of plastic may swell if exposed to organic solvents. Plastics also have low abrasion resistance. Use solid plastic nozzles only with selected pesticides. Some plastic nozzles have stainless steel orifice inserts, making them much more resistant to wear. The inserts also reduce swelling problems.
- Tungsten carbide and ceramic nozzles are highly resistant to abrasion and corrosion. To reduce costs, manufacturers use tungsten carbide or ceramic inserts with brass or plastic nozzle bodies. Use these types of nozzles for high-pressure and abrasive sprays. Ceramic nozzles are usually affordable and last a long time, but they are fragile and so may crack if overtightened.

List the important factors to consider when selecting nozzles for a given application.

FIGURE 9-8.

Most pesticides are applied with a nozzle type designed to produce a tapered, flat-spray pattern, like this one.

Selecting Nozzles

Picking the best nozzle for your particular application method, target pest, site conditions, and equipment can increase the safety and effectiveness of your pest control efforts. This section explains the various nozzles used in agricultural settings and the methods used to select the right nozzle for the job.

The spray nozzle you choose directly affects droplet size, spray uniformity, spray coverage, and drift potential, which in turn impacts pest control, economics, and environmental quality. Although nozzles have been developed for practically every kind of spray application, only a few nozzle types are commonly used to apply pesticides and fertilizer-pesticide combinations. Most pesticides used in ground applications (spraying row crops or weeds) are applied with a nozzle type designed to produce a tapered, flat-spray pattern (Fig. 9-8). Table 9-1 compares a sampling of popular spray tips used to apply pesticides in agricultural areas and illustrates their spray patterns. Figure 9-9 shows the color codes that represent standardized droplet sizes used by nozzle manufacturers to make nozzle selection easier.

Color codes for droplet sizes

Category	Symbol	Color code	Droplet size range (microns)
Extremely fine	XF	Purple	< 60
Very fine	VF	Red	60–145
Fine	F	Orange	146–225
Medium	M	Yellow	226–325
Coarse	C	Blue	326–400
Very coarse	VC	Green	401–500
Extremely coarse	EC	White	501–650
Ultra coarse	UC	Black	> 650

FIGURE 9-9.

Droplet-size categories for nozzles with symbols and color codes.

TABLE 9-1:

Common nozzles and their uses

Nozzle type	Description	Spray pattern illustration	Recommended uses
		Flat-fan	
even	Fan-shaped pattern with even distribution of droplets across width of fan.		Use to apply preemergence and postemergence herbicides, insecticides, and fungicides. Use at pressures of 20–40 psi; keep pressure as low as possible when spraying weeds. Use on a boom when applying separate bands of spray that should not overlap.
off-center	Fan-shaped pattern with angle to one side.		Use for herbicide applications to orchard or vineyard soil on both sides of the plant row. Use on ends of spray booms to extend the spray pattern. Use at pressures of 20–40 psi; keep pressure as low as possible when spraying weeds. Requires 100% overlap.
low-pressure	Fan-shaped pattern with fewer droplets at sides than in center of pattern.		Use to apply preemergence and postemergence herbicides, insecticides, and fungicides. Use at pressures of 20–60 psi; keep pressure as low as possible when spraying weeds. Suitable for overlapping with other nozzles to produce a wide spray swath.
extended-range (broadcast)	Wide, fan-shaped pattern ranging from fine to coarse droplets.		Use for soil and foliar applications when better coverage is required than can be gotten from flooding or turbo flooding nozzles. Best suited for use with electronic controllers that will control the spray rate either by adjusting the spray pressure or pulse width modulation. Use at pressures of 10–30 psi for soil applications and 30–60 psi for foliar applications (pressures above 25 psi can increase the likelihood of drift).
		Cone	
hollow	Hollow cone pattern of fine droplets at angles ranging from 20–110 degrees.		Use to apply postemergence contact herbicides, contact fungicides, and contact insecticides in dense foliage. Often used with air blast sprayers. Use at pressures of 40–120 psi.

TABLE 9-1:

Common nozzles and their uses (continued)

Nozzle type	Description	Spray pattern illustration	Recommended uses
solid	Solid cone pattern of large droplets at angles ranging from 20–110 degrees.		Use to apply soil-incorporated and preemergence herbicides. Use when you want heavier droplets to reduce drift or you require larger volumes to ensure complete coverage. Use at pressures of 40–120 psi.
		Turbulence chamber (turbo)	
flood	Wide, fan-shaped pattern of coarse droplets.		Use to apply postemergence systemic, soil-incorporated, and preemergence herbicides, systemic fungicides, and systemic insecticides. Requires at least 50% overlap for proper application uniformity. Use on a boom to apply pesticides to soil and when you need to reduce drift. Use at pressures of 5–20 psi.
flat	Wide, flat, tapered fan-shaped pattern of large droplets.		Originally designed for use in the application of postemergence products, but can be used in any application to reduce drift. Use with automatic sprayer controllers. Requires 50 to 60% overlap to achieve uniform application across the boom. Use at pressures of 15–100 psi.
		Other	
solid stream	Low- or high-pressure solid stream; high pressure breaks spray into fine to medium droplets.		Use on booms in row crops to apply all types of pesticides in bands. Use at pressures of 5–200 psi.
flooding	Wide, fan-shaped pattern of coarse droplets.		Use to apply soil-incorporated herbicides and mixtures of herbicides and liquid fertilizers. Use on a boom when you need a high volume of liquid and drift is a concern. Use at pressures of 5–20 psi.
broadcast	Wide, fan-shaped pattern ranging from fine to coarse droplets.	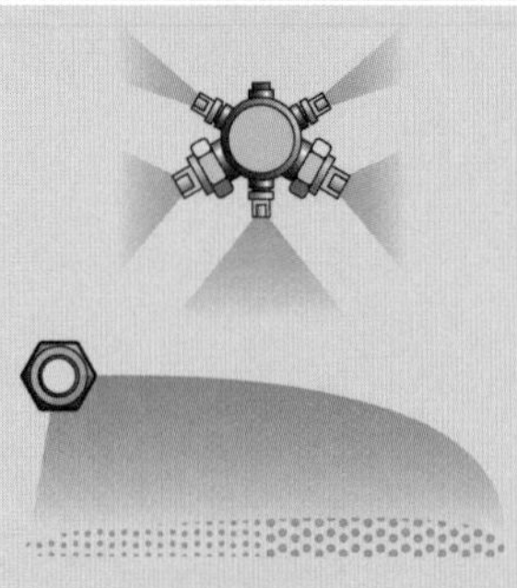	Use on boomless sprayers to deliver a wide (30–60 foot) swath in places where a boom is not practical, like in end-row, pasture, and orchard spraying. Use for herbicide applications to control weeds and brush in pastures or rangelands. Use at pressures of 10–30 psi.
air-injection/ air-induction/ Venturi	Fan-shaped pattern of coarse droplets. Can also have a hollow cone pattern.		Use on a boom for high-pressure applications when drift needs to be reduced. Use to control broadleaf weeds. Use at pressures of 40–50 psi and higher.

Application Methods and Equipment

List the types of application equipment, and describe the advantages and limitations of each type.

List the types of application equipment used to apply liquids, and describe the situations in which each should be used.

List the types of application equipment used to apply dusts, and describe the situations in which each should be used.

List the types of application equipment used to apply granules, and describe the situations in which each should be used.

List types of bait application equipment, and explain how they work.

Describe how to maintain and clean different kinds of equipment.

Table 9-2 describes various application methods, such as spot, banded, and broadcast treatments. The situation in which you might use a particular method, the method's benefits and drawbacks, and the equipment used to make such applications are also included.

Maintaining Application Equipment

Effective pesticide application depends on properly maintained and adjusted application equipment. Regular inspections and periodic maintenance programs help you avoid accidents or spills caused by ruptured hoses, faulty fittings, damaged tanks, or other problems.

Inspect application equipment for wear, corrosion, or damage before each use. Replace or repair faulty components. Thoroughly clean equipment after every application. Wear PPE, including rubber gloves and eye protection, when cleaning or repairing the equipment. When not in use, store equipment in a way that prevents deterioration or damage.

Liquid Application Equipment

Preventing Problems

Take the following preventive steps to reduce problems of sprayer malfunction or breakdown and to maintain uniform and accurate application.

Use clean water. Water that contains sand or silt causes rapid pump wear and can clog screens and nozzles. Whenever possible, use water pumped directly from a well, and make sure all filling hoses and pipes are clean. Always filter water before putting it into the sprayer tank, and have your fill station on the downstream side of your filtering system. Also, measure the pH of the water to be sure it is adequate for the intended pesticide use. (See Sidebar 11-2, which describes how to check and adjust the water's pH.)

Keep screens in place. Filter screens remove foreign particles from the spray liquid. It is a nuisance to remove collected debris from the screens, but debris accumulation indicates that the screens are doing their job (Fig. 9-10). Removing screens because they keep plugging only increases wear on pumps and nozzles, and can result in plugged nozzles, which makes applications inaccurate. Make sure screens are the proper size for the type of pesticide being applied. If excessive plugging does occur, try to eliminate the cause, for example, by changing water sources and cleaning the nozzles.

FIGURE 9-10.

Clean screens often to prevent clogging in nozzles and in hoses running to and from pumps.

TABLE 9-2:

Pesticide application methods and equipment commonly used in agricultural settings

Application method	Situations	Benefits/drawbacks	Equipment
band and directed spray applications	Treatment of row crops to manage insects, diseases, and weeds. Treatment of noncrop agricultural areas and grasslands to prevent the spread of weeds. Treatment of various pests of livestock and poultry. Treatment of weeds in orchards and vineyards.	Benefits: Uses less pesticide than other methods, so reduces cost per treatment. Can be targeted very effectively as directed spray, shielded spray, or hooded spray applications. Can be used at different times in a crop's life cycle. Drawbacks: Requires special equipment to make applications more targeted. The nozzles on some systems are arranged with drop hoses or thin metal strips that can twist and rotate the nozzles as the applicator moves through the field. This accidental twisting can decrease spray pattern uniformity and may cause drift.	hydraulic (liquid) sprayers low-pressure boom sprayers front-mounted boom sprayers granule spreaders spinning disc sprayers air shear sprayers
spot treatments, spray-to-wet treatments, drizzle treatments, pour-on treatments	Early treatment of mite and insect infestations that are concentrated in just a few areas, and have yet to spread to other parts of a field. Treatment of patches of weeds scattered throughout a field, or weeds that are taller than crop/desired plants. Treatment of field edges with herbicide or insecticide to prevent field infestation. Sometimes used to apply fungicides to limited areas. Treatment of poultry and livestock for a variety of pests.	Benefits: Small, hand-carried sprayers can reach areas that tractor-mounted or self-propelled sprayers cannot reach. Small sprayers keep pesticides on target and reduce environmental contamination when calibrated and used correctly. Can reduce the amount of pesticide required to control pests from 70 to 90% over broadcast applications. Drawbacks: Backpack (knapsack) sprayers are tiring to carry and operate, and they can treat only a small area. Walking pace for carried sprayers must remain steady to ensure proper coverage, even over small areas. A steady walking pace may be hard to maintain, depending on the terrain. You cannot control the spray volume delivered from spot to spot, so some areas may receive more material than necessary, and some may receive less.	syringe sprayers hand-operated backpack or knapsack sprayers (lever-operated, trigger pump, mistblowing, hose-end, push-pull hand pump, compressed-air) nonpowered and powered wheelbarrow sprayers powered backpack sprayers estate sprayers rope wick or canvas wiper applicators drizzle applicators trigger-release spray guns
precision (patch) spraying, spray-to-wet treatments, drizzle treatments	Early treatment of mite and insect infestations that are concentrated in just a few areas, and have yet to spread to other parts of the field. Treatment of patches of weeds scattered throughout a field, or weeds that are taller than crop/desired plants. Treatment of field edges with herbicide or insecticide to prevent field infestation. Sometimes used to apply fungicides to limited areas. Treatment of cattle and some poultry for a variety of pests.	Benefits: For sprayers with GPS and rate controllers installed, pesticides can be targeted accurately to precise areas. Can reduce the amount of pesticide required to control pests from 70 to 90% over broadcast applications. Drawbacks: System controllers can be expensive up front, and they require the operator to understand how to program such systems. These sprayers require more maintenance over their lifetime than other types of sprayers.	boom sprayers equipped with VRA systems spinning disc sprayers drizzle applicators ultra-low-volume sprayers

TABLE 9-2:

Pesticide application methods and equipment commonly used in agricultural settings (continued)

Application method	Situations	Benefits/drawbacks	Equipment
basal	Treatment of weeds growing around the base of established trees and vines. Treatment of insects infesting the base of established trees and vines. Removal of problem woody plants that are less than 8 inches in diameter, or that have very thin bark.	Benefits: Pesticide applied to bark for removal of woody pest plants moves systemically throughout the plant to kill it. Causes little or no damage to surrounding plants when spray is accurately targeted. Can be used for weed and insect control in vineyards and orchards where plants are well-established, allowing the applicator to precisely target pesticides to affected areas and reduce the amount of pesticide needed to control pests. Drawbacks: Treatments cannot be made in rainy weather or if rainy weather is predicted. People and animals are easily exposed to pesticides on treated bark in areas of animal and human activity. Protection activity may be short-lived, depending on the product applied. The effectiveness of treatment depends on the temperature during and just after application, as well as the age of the plant and corkiness /thickness of the bark.	syringe sprayers trigger pump sprayers low-volume backpack sprayers wick applicators
broadcast	Treatment of insects and disease in areas of dense foliage. Treatment of pests in orchards. Treatment of large areas where there are numerous insects or weeds (a high density of pests). Used in pastures and grasslands when weeds have crowded out sensitive desirable plants.	Benefits: Provides good penetration and coverage of plant surfaces and animal hair, especially in dense foliage and when animal hair is thick. Large tanks on some equipment allow many acres to be treated in just one application. Can be used in many different situations. Drawbacks: May cause drift hazards if droplet size is too small for conditions at the site. Conditions at the application site can cause uneven deposit of pesticide. Pests are not usually uniformly distributed throughout the field, so making this type of application may waste some pesticide. Uses a lot of water, power, and fuel.	backpack mistblowers air-sleeve sprayers tunnel sprayers (using hydraulic nozzles) air blast sprayers orchard sprayers using hydraulic nozzles, air shear nozzles (on machines with centrifugal fans), or rotary nozzles (mounted in front of propeller fans) oscillating boom sprayers high-pressure hydraulic sprayers air-assist sprayers electromagnetic sprayers spreaders tractor-mounted airflow granular applicators power dusters
rope-wick or canvas-type wiper, self-applicator setups	Treatment of tall weeds in fields and pastures, especially where drift is a concern. Treatment of external animal parasites.	Benefits: Equipment reduces drift by targeting herbicides extremely accurately—the pesticide is deposited directly onto pest plants. Wick applicators allow the selective application of broad-spectrum pesticides. Animals are automatically treated when moving through areas set up with self-applicators. Drawbacks: Rope wicks must be watched carefully to avoid oversaturation and dripping or having too dry a wick. Wicks can accumulate dirt on their surfaces, keeping pesticide from reaching target plants. Weedy species must be taller than desirable plants. Most animals must be trained to accept the devices.	rope wick applicators canvas wipers enclosed boom rotating wipers (All of these can be on a boom attached to an ATV or tractor. Booms can also be hand-carried.) face and back rubbers cables ropes dust bags dust boxes

TABLE 9-2:

Pesticide application methods and equipment commonly used in agricultural settings (continued)

Application method	Situations	Benefits/drawbacks	Equipment
dip, pour-on, spray-dip	Treatment for root disease prior to transplanting. Treatment for nematodes prior to transplanting. Treatment for insects that affect roots prior to transplanting. Treatment of livestock for external parasites.	Benefits: Prevents the spread of pests through soil contamination caused by transplanted crops. Systems can be mechanized to reduce exposure risks for workers and increase efficiency. Controlled application means pesticide is targeted very accurately to affected parts of the plant or animal. Immersed animals are fully and evenly treated. Pour-on treatments are convenient and lower worker exposure risks. Drawbacks: Can be expensive and time consuming. Not all dip treatments have been proven to work effectively. Exposure risk is elevated when systems are not mechanized.	dipping vats mechanical systems that work with pesticide vats to dip plants or animals spray-dip machines
foliar	Used in insecticide and fungicide applications on crops. Used in herbicide applications after weeds have emerged (best to use when they are small and actively growing, when more established as in the early bloom stage, during active growth in the fall, or when weeds are taller than crop plants).	Benefits: Applied directly to affected area of plants. Flexible—can treat many different problems during a variety of stages of pest development. Drawbacks: Depending on the application method, drift can be a problem. Can affect nontarget species. Runoff can become a problem.	coldfoggers air-assisted electrostatic handguns pulsefoggers low-volume hydraulic sprayers high-volume hydraulic sprayers electrostatic sprayers rope wick and canvas wiper applicators power dusters rear-mounted boom sprayers tunnel sprayers drizzle applicators
soil application: drench furrow soil incorporation	Treatment of weeds, insects, nematodes, or pathogens prior to or during planting. Treatment of weeds, insects, or pathogens before and after planting.	Benefits: Can be used to treat many different pest problems at various stages of development. Can be used to prevent problems with nematodes and pathogens—may be the only viable way to get rid of nematode pests, especially in tree and vine crops. Drench treatments are highly targeted to the plants or trees that are affected by the pest. Furrow treatments are precisely targeted and use low volumes of pesticide, reducing environmental hazards. Soil-incorporated pesticides mix thoroughly with soil so are not as prone to drift (though leaching can be a problem—always read the label for precautions). Drawbacks: In soil-incorporation treatments, pesticide must be evenly distributed within the soil in order for it to reach the target pest or be taken up by the roots of plants as they grow. Drench treatments can be time, water, and labor intensive. Leaching can be a problem, especially in sandy soils.	low-volume microtube in-furrow sprayers vehicle-mounted or towed toolbar fitted with chisel tines that allow for the mixing of pesticide into the soil as the application proceeds granule spreaders low-volume boom sprayers

TABLE 9-2:

Pesticide application methods and equipment commonly used in agricultural settings (continued)

Application method	Situations	Benefits/drawbacks	Equipment
soil injection	Treatment of nematodes, insects, weeds and pathogens prior to planting.	Benefits: Addresses problems before crop is planted, which is especially important for long-term crops like trees or vines. Tines along the toolbar can be adjusted to penetrate the soil at varying depths, so the system is flexible. Drawbacks: This method is expensive. Many chemicals used in soil injection are environmentally problematic and require restricted materials permits and special pesticide licenses or certifications for applicators.	vehicle-mounted or towed toolbar fitted with chisel tines
chemigation Describe the situations in which chemigation systems can be used.	Treatment of weeds, insects, or pathogens to large areas before and after planting.	Benefits: Can use existing irrigation equipment to apply pesticides. Very efficient. Uses precise amounts of pesticides metered by control systems. Drawbacks: Specialized equipment must be purchased and set up, which can be expensive. Systems are complex and must meet all state and local regulatory requirements for the protection of groundwater. Can result in uneven applications if system is not maintained adequately or if rolling or uneven soils are treated.	Setup depends on irrigation systems available at the site.
bait delivery	Control of vertebrate and invertebrate pests.	Benefits: Can keep nontarget organisms out, depending on the design. Can be placed out of reach of nontarget organisms, pets, and children. Uses only the amount of pesticide needed to control the targeted pest. Drawbacks: Specialized equipment must be purchased and set up, which can be expensive.	hand-operated injectors mechanical bait applicators (burrowing devices) bait stations

Use chemicals that are compatible with the sprayer and pump. Spray chemicals are corrosive to some metals and can speed the breakdown of rubber and plastic components. Recognize limitations in existing spray equipment. Avoid problems by modifying the equipment to accommodate the corrosive pesticides. Otherwise, use the equipment only for chemicals that are not corrosive. Sometimes it is possible to replace parts of a sprayer with corrosion-resistant materials.

Properly clean nozzles. Spray nozzles are made to precise specifications. Never use any metal object to clean or remove debris. These may damage the orifice, adversely changing the spray pattern and spray volume. Clean nozzles by flushing with clean water or a detergent solution. Remove stuck particles with a soft brush. Nozzle suppliers sell special brushes for this purpose. Always wear rubber gloves when handling or cleaning spray nozzles. Never blow through them with your mouth, because nozzles usually contain pesticide residues. Use an air compressor if needed, but protect your eyes and skin (Fig. 9-11).

Flush sprayers before use. Use clean water to flush new sprayers and sprayers coming out of storage. Flushing removes foreign particles, dirt, and other debris. The manufacturing process may leave metallic chips, dirt, or other residue in the tank or pump. Storage always subjects spraying equipment to the possibility of being contaminated with dirt, leaves, rodent debris, and rust.

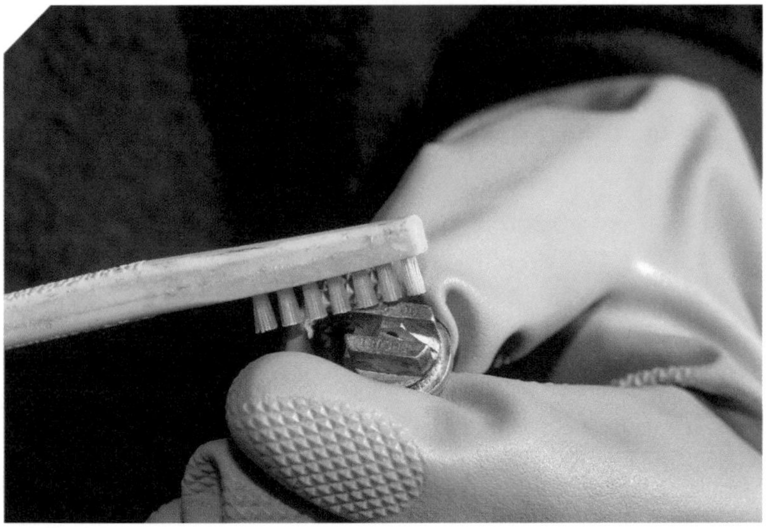

FIGURE 9-11.

To clean a clogged nozzle, use compressed air or water for flushing the orifice. Never put your mouth to a nozzle. Use a soft brush to remove stuck objects. Do not use any type of metal device to remove debris, because you may damage the orifice.

Clean sprayer after use. Cleaning spray equipment at the end of each job is important. Clean your sprayer according to the instructions on the label of the last pesticide used. Following the label's directions will ensure that your efforts will remove most residues that can contaminate future tank mixes or damage crops or treated surfaces. Avoid leaving pesticide mixtures in a sprayer overnight or for longer periods of time. Prolonged contact increases chances of corrosion or deterioration of sprayer components. Some pesticides settle out and may be hard to get back into suspension after being left in an idle sprayer, causing plugging problems. After mixing them with water, certain pesticides lose their effectiveness quickly. Finally, pesticides left in an unattended sprayer may present a hazard to people, wildlife, or the environment.

If possible, apply leftover spray material to an appropriate (registered) target site. Otherwise, treat unused pesticide mixtures as hazardous wastes. Additionally, some pesticides are hard to remove entirely from the tank, and you will need to use special cleaners to thoroughly remove their residues. Check the pesticide label, because it may recommend the use of specific cleaners. You may want to use separate tanks if you apply a hard-to-clean pesticide, especially when using herbicides.

Clean the sprayer and flush out the tank in the field at the application site whenever possible. If this is not possible, contain the wash water and use it for mixing other pesticides of the same type. Remember that repeated cleaning in a particular location can result in contamination of that area unless you carefully contain wash water. If you cannot use the contaminated water in your next tank mix, have it transported to a landfill or waste unit that can accept hazardous waste (known as a Class I disposal site). Never drain rinse water onto the ground or into sewer or septic lines. If you will be using a different pesticide in the sprayer next time you use it, you will need to do a more thorough cleaning. Check the previously used pesticide's label for recommended tank cleaning agents.

Inspection and Maintenance

Perform regular inspections and periodic maintenance on spraying equipment to keep it in good operating condition. Take care of simple maintenance, such as greasing bearings and drive lines, while inspecting the equipment. Always check for the following problems:

- weakened hoses
- leaking fittings
- damage to the tank or tank protective coating
- broken regulators or pressure gauges
- worn nozzles
- worn bearings
- damaged tires (if equipped)
- other mechanical defects or wear

Equipment with a self-contained engine requires additional maintenance. Check oil and water levels regularly. Change air filters, oil filters, and motor oil according to the manufacturer's recommendations. Clean and service batteries.

Dust and Granule Applicators

Thoroughly clean dust and granule applicators after each use. Be sure to remove all pesticides and pesticide residue. Once clean, lubricate chains, auger bearings, and other moving parts according to the manufacturer's instructions. Inspect the equipment for wear and corrosion. Repair rusted or corroded areas to prevent them from getting worse.

Dust Applicators

Before using a dust applicator, inspect it carefully. Check the inside of the bag or chamber to make sure it is dry and free of residue or potential obstructions. Next, inspect the screw threads where the nozzle or application tip is attached and make sure they are clean. Then make sure that the application tip is securely attached. Finally, inspect the application tips and extensions for cracks and replace any damaged or worn accessories. Dispose of worn or broken dust applicators as you would any other pesticide-contaminated item.

After use, clean the equipment with a nylon brush using soap and water. Make sure parts that contact dry pesticide are thoroughly and completely dry before reuse. Carefully inspect the cap, screw threads, and application tubes for cracks and other parts for damage or obvious wear. If you find any flaws (such as cracks in plastic or corrosion in metal parts), remove and replace the worn parts before using the applicator again.

Granule Applicators

Before using granule applicators, inspect them for wear and to make sure there are no pesticide residues left in the hopper or anywhere else on the machine. Remove and replace worn or damaged parts, and check to be sure there are no blockages before loading the equipment. Check to make sure the gear cover is in place on applicators that have one; this helps protect moving parts that are easily damaged by dirt or pesticide residue.

Empty and thoroughly wash granule applicators after every use. Normally, cold water is all that you will need to remove pesticide residue from the equipment; however, some pesticides will require scrubbing or the use of hot water to loosen built-up residues. It may be useful to close the spreader so the hopper can be filled completely with water and then drained. If you have to use abrasive cleaners, be careful not to damage the equipment as you work.

Lubricate granule applicators only if the manufacturer recommends it. Be careful of too much lubrication, because grease can increase buildup of pesticide residues and dirt that can damage an applicator's moving parts. Read the manufacturer's recommendations carefully to find out if your equipment requires lubrication before taking this step.

Maintaining Chemigation Equipment

Periodic monitoring of chemigation systems can help you ensure that they are operating safely and effectively. The following items should be inspected thoroughly before you begin chemigation activities:

- main pipeline check valve
- vacuum relief valve
- low-pressure drain
- chemical injection line check valve
- main control panel for the irrigation system and pumping plant
- chemical injection pump safety interlock
- injection system (inline strainer, manual valve, and chemical storage tank)
- irrigation pump
- injection pump
- power source

Repair or replace any parts you find that are damaged or worn. Be sure to recalibrate the system any time maintenance has been performed and parts have been replaced.

Flushing Irrigation and Injection Systems

Chemigation equipment must be thoroughly flushed after each use in order to ensure safe operation during the next application. After completing an application, run the irrigation system for at least 10 minutes to flush out any chemicals that may remain. You may have to run the system for more than 10 minutes if you are using drip irrigation, as water takes longer to run through low-volume systems. For any irrigation system type that shuts down automatically at the end of an application, be sure to flush it as soon as possible after shutdown is complete. You must also flush systems that have been shut down because of a malfunction or loss of water pressure. Do this as soon as you can after discovering the shutdown. In both of these situations, it is best to flush the system for at least 30 minutes, just to be sure all traces of the pesticide have been run out of the system.

Use clean water to flush your injection system after each use to prevent pesticides from accumulating. It is best to flush the injection system as you are irrigating, so that whatever pesticides you flush out are applied to the same site.

Chapter 9 Review Questions

1. Match the application method with its major drawback.

1. band and directed spray applications	a. leaching can be a problem, especially in sandy soils
2. broadcast	b. uses a lot of water, power, and fuel
3. dip or spray-dip applications	c. requires special equipment to make applications more targeted
4. soil incorporation	d. requires specialized equipment and setup to meet regulatory requirements for protection of groundwater
5. chemigation	e. exposure risk is higher when systems are not mechanized

2. True or false?

☐ True ☐ False a. Piston pumps are the best pumps to use for spraying abrasive formulations.

☐ True ☐ False b. Nozzles with larger orifices wear more slowly than nozzles with smaller orifices.

☐ True ☐ False c. Fiberglass tanks must be checked carefully for scratches or abrasions, since they absorb pesticides that can contaminate the next tank mix.

☐ True ☐ False d. The most expensive nozzles are made of brass.

☐ True ☐ False e. If spray pressure is dropping and there is unusual strain on the pump, it is most likely because the filter screens are clogged.

☐ True ☐ False f. You can safely remove particles stuck in a nozzle using a thin copper wire.

☐ True ☐ False g. Cleaning a sprayer repeatedly in a particular location can result in contamination unless you carefully contain the wash water.

☐ True ☐ False h. It takes longer to flush a drip irrigation system after chemigation than other types of irrigation systems.

3. Match the nozzle with its spray pattern and uses.

1. low-pressure flat-fan nozzles	a. These nozzles produce a fan-shaped pattern of coarse droplets. They are used on a boom to apply herbicides in situations where drift needs to be reduced.
2. even flat-fan nozzles	b. These nozzles create more spray droplets in the center and fewer droplets on the side so that the pattern tapers off at each end. They are used with soil-applied herbicides, fungicides, and insecticides.
3. solid cone nozzles	c. These nozzles produce an even distribution of droplets in a fan-shaped pattern. They are used when you don't want the herbicide, fungicide, or insecticide spray to overlap.
4. air-injection/air-induction/Venturi nozzles	d. These nozzles are used to apply large volumes of soil-incorporated and preemergence herbicides. They produce large droplets that help reduce drift.
5. solid stream nozzles	e. These nozzles are used on booms in row crops to apply all types of pesticides in bands at pressures ranging from 5-200 psi.

4. If you often spray highly abrasive formulations, you should use nozzles made out of ________. Select all that apply.

- ☐ a. aluminum and monel
- ☐ b. ceramic
- ☐ c. solid plastic
- ☐ d. stainless steel
- ☐ e. tungsten carbide
- ☐ f. brass

5. Match the application situation with its appropriate liquid applicator.

1. You want to control tall weeds in fields or pastures, and drift is a major concern.	a. spray-dip machine
2. You want to control insects or disease in areas of dense foliage.	b. low-volume boom sprayer
3. You want to control external parasites on livestock that must be evenly treated.	c. wick applicator
4. You need to control nematodes in tree or vine crops.	d. air blast sprayer

Chapter 10
Calibrating Pesticide Application Equipment

Why Calibration Is Essential 128
Equipment Calibration Methods 128
Calculation for Active Ingredient, Percentage Solutions, and Parts per Million Dilutions 146
Using System Monitors and Controllers 149
Chapter 10 Review Questions 151

Knowledge Expectations

1. Define calibration, and explain why accurate calibration is essential to safe, effective pest control.
2. List the tools needed for calibration activities.
3. List the variables that must be measured to calibrate a sprayer.
4. Describe how to calibrate liquid sprayers, and be able to calculate speed, gallons/ minute, and nozzle output using formulas.
5. Describe methods used to determine how much pesticide to put into the hopper or tank for a specific application rate over the total area of the application site.
6. Describe the best way to change the output of various pesticide application equipment and the consequences of each change.
7. Describe how to calibrate dry applicators.
8. Describe how to determine the correct amount of pesticide needed for a particular application, including diluting a pesticide correctly.
9. Be able to calculate the active ingredient concentration of pesticides using formulas.
10. Explain how system controllers can impact the calibration of equipment and calculations necessary to apply pesticides effectively.
11. Explain the importance of properly calibrating sensors that are part of a system controller.

Why Calibration Is Essential

Define calibration, and explain why accurate calibration is essential to safe, effective pest control.

The term *calibration* refers to all the adjustments you make to be sure you apply the correct amount of pesticide to the treatment area. The main reasons for calibrating application equipment are to figure out how much pesticide to put into the tank or hopper and what the application rate should be. Calibration is necessary for

- controlling pests effectively
- protecting human health, the environment, and treated surfaces
- making efficient, effective applications
- determining spray volume
- applying pesticides at legal levels

Effective Pest Control. Manufacturers of pesticides spend millions of dollars researching their products. Their research includes determining the correct amount of pesticide to apply to effectively control target pests. Calibrating your equipment before each application helps to ensure that the proper amount of pesticide will be effectively applied. Applying the correct amount helps to ensure effective pest control and minimize the chance of resistance occurring in target pest populations. If the application is ineffective due to poor calibration, then it may need to be reapplied or crop yield may go down.

Environmental Concerns. Poor calibration of pesticide application equipment may cause environmental problems. Calibrating equipment to maintain application rates within label requirements helps protect people, livestock, beneficial insects, and wildlife. It also reduces the potential for contaminating surface water, groundwater, and the air.

Protecting Treated Surfaces. Certain pesticides can damage treated surfaces when used at higher-than-label rates, causing problems like phytotoxicity (plant injury), staining, or corrosion. Manufacturers evaluate these potential problems while testing their chemicals to determine safe concentrations. Calibrating your equipment helps keep you from using damaging amounts of pesticide during your application.

Preventing Waste of Resources. Using the improper amount of pesticide wastes the pesticide, time, and money. Proper calibration helps you use less fuel and reduces labor costs and equipment wear and tear.

List the tools needed for calibration activities.

Legal Concerns. Applicators who do not properly calibrate equipment may end up using pesticides improperly and therefore may be subject to criminal and civil charges, resulting in fines, jail time, and lawsuits. Applicators are liable for injuries or damage caused by improper pesticide application, so they must take care to calibrate equipment before any application.

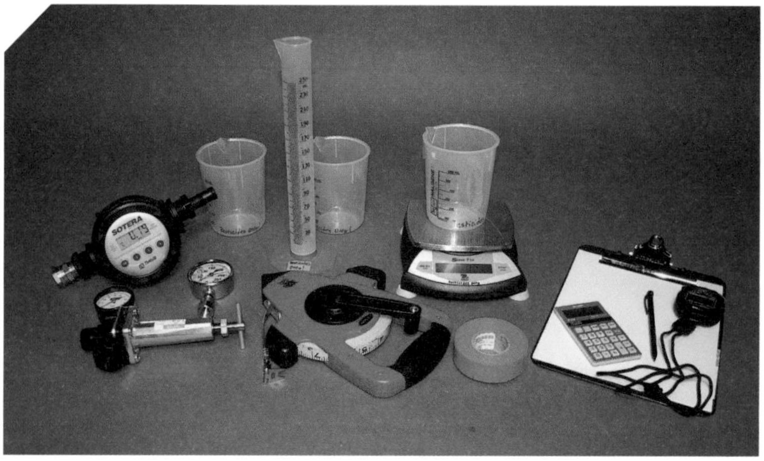

FIGURE 10-1.

A few simple tools are required for calibrating a pesticide sprayer. These include a stopwatch, measuring tape, several calibrated containers, a scale, pocket calculator, pressure gauge, flowmeter, and flagging tape.

Equipment Calibration Methods

Sidebar 10-1 lists the tools you need to calibrate pesticide application equipment. Put these items in a small toolbox and use them only for calibrating your equipment (Fig. 10-1). Keep your tools clean and in good working condition; make equipment calibration a professional operation. Liquid application equipment and dust or granular application equipment require different calibration methods.

NOTE: Pesticide application equipment and the discharge from application equipment being calibrated may contain pesticide residue.

SIDEBAR 10-1

Tools Needed for Calibration

1. **Stopwatch.** Use a stopwatch for timing travel speed and flow rates. Never rely on a wristwatch unless it has a stopwatch function.
2. **Measuring tape.** Use a 100-foot moisture- and stretch-resistant measuring tape for marking off the distance to be traveled and measuring spray swath width.
3. **Calibrated container.** Use a 1- or 2-quart container, calibrated for liquid ounces, for measuring spray nozzle output.
4. **Scale.** Use a small scale capable of measuring pounds and ounces for weighing granules collected from a granule applicator. The most accurate weight measurements come from scales that have maximum capacities from 5 to 10 pounds.
5. **Pocket calculator.** Use a pocket calculator for making calculations in the field.
6. **Pressure gauge.** Use an accurate, calibrated pressure gauge that has fittings compatible with spray nozzle fittings for checking boom pressure and for calibrating the sprayer pressure gauge.
7. **Flowmeter.** Use a flowmeter attached to a flexible hose or filling pipe for measuring the amount of water put into a tank. You can also use this device for measuring tank capacity and for determining the amount of liquid used during a calibration run. Both mechanical and electronic flowmeters are available. If these are not available, a calibrated 5-gallon pail can be used instead.
8. **Flagging tape.** Use colored plastic flagging tape for marking off measured distances when determining applicator speed.

List the variables that must be measured to calibrate a sprayer.

Describe how to calibrate liquid sprayers, and be able to calculate speed, gallons/minute, and nozzle output using formulas.

Always wear chemical-resistant gloves and other personal protective equipment (PPE) to prevent pesticide contamination. Read Chapter 7 for information on selecting the proper PPE.

Calibrating Liquid Sprayers

In order to properly calibrate liquid sprayers, you need to measure these four factors:

- tank capacity
- travel speed
- flow rate
- spray swath width

Before making any calibration measurements, be sure to service the sprayer as described in Chapter 9. Once the sprayer is in good working condition, use the following basic formula for calibration:

Spray Volume (gpa) =
Flow Rate (gpm)/Land Rate (Land Rate is Speed × Swath Width)

Formulas for determining gallons per acre (gpa), flow rates (in gallons per minute, or gpm), speed, and swath width can be found in sidebars throughout this chapter.

Tank Capacity. Physically measure the capacity of the spray tank (or tanks, if the equipment has more than one). Never rely on manufacturers' tank size ratings. They may be estimated or may not take into account fittings installed inside the tank. Also, the capacity of spray lines, pumps, and filters influences tank volume. To accurately calibrate your equipment, you need to know exactly how much liquid the spray tank holds.

FIGURE 10-2.

Flowmeters, similar to the one shown, can be used to measure the volume of spray tanks.

Put the sprayer on a perfectly level surface. Be sure the tank is completely empty, then close all valves to prevent water leaks. Add measured amounts of clean water until you completely fill the tank. Use a flowmeter attached to a hose (Fig. 10-2) or a bucket or other container of known volume. A 5-gallon bucket works well for smaller sprayers. Be sure to calibrate and mark the bucket before using it to fill the tank. If you are not using a flowmeter, use smaller-volume calibrated containers to top off the tank. Record the total volume of water you put into the tank. Paint or engrave this figure onto the outside of the tank for permanent reference.

While filling the tank, calibrate the tank's sight gauge. Make marks on the tank or gauge as you put in measured volumes of water. If the unit does not have a sight gauge, mark volume increments on a dipstick. Then, always keep this dipstick with the tank. Use 1-gallon marks for tanks with a capacity of 10 gallons or less. Use increments of 5 or 10 gallons for tanks having a total capacity of 50 gallons or less. On larger tanks, use increments of 10 to 20 gallons. Once you calibrate the sight gauge or dipstick, you can measure how much liquid is in the tank when it is not entirely full. Always return tanks to a level surface when reading the sight gauge or dipstick.

Travel Speed. Always measure travel speed under actual working conditions because tire slippage will vary on slopes and with different ground surfaces. For instance, to calibrate an orchard sprayer, fill the tank at least halfway with water and take it to the orchard. Calibrate row crop and field sprayers in the fields you plan to treat. Tractors travel faster on paved or smooth surfaces than on soft dirt or clods. Never rely on tractor speedometers for mile-per-hour measurements. Tractor wheel slippage and variation in tire size produce as much as a 30% difference in actual versus indicated speed. When calibrating a backpack or handheld sprayer, walk on ground that is similar to the area you plan to spray.

Using a l00-foot tape, measure off any convenient distance. It can be more or less than 100 feet, but calibration accuracy increases if you use longer distances (from 200 to 300 feet). Sometimes multiples of 88 feet are chosen because 88 feet is the distance covered in 1 minute while traveling 1 mile per hour. In orchards or vineyards, a given number of tree or vine spaces of known length provides a useful reference. Indicate the beginning and end of the measured distance with colored flagging tape.

FIGURE 10-3.

Measure and mark off a known distance when calculating the speed of travel of the application equipment. Use a stopwatch to time the travel of the sprayer through the measured distance.

Have someone drive (or walk, if calibrating a backpack sprayer) through the measured distance. Maintain the speed desired for an actual application. Choose a speed within a range appropriate for the application equipment, plant size, and amount and density of foliage. Younger plants with smaller leaves may allow a faster travel speed than a mature plant (orchard or vineyard) or a plant with dense foliage. You should stay within the appropriate speed range of the equipment after accounting for plant size and foliage. When using a tractor, note the throttle setting, gear, and tractor engine speed (rpm). Be sure to bring the equipment up to the actual application speed before crossing the first marker. Use a stopwatch to determine the time, in minutes and seconds, required to traverse the measured distance (Fig. 10-3). For best results, repeat this process two or three times and take an average. Follow the procedure in Sidebar 10-2 to calculate the actual speed of the equipment. You can also use a GPS unit to check the measurements you take, using the flagging tape and stopwatch to ensure accuracy.

Flow Rate. Measure the actual output of the sprayer when nozzles are new, then periodically check for nozzle wear thereafter. Manufacturers print charts showing output of given nozzle sizes at specified sprayer pressures. However, you should check output under actual conditions of operation. Manufacturer's charts are most accurate when using new nozzles.

Used nozzles may have different output rates because of wear. However, even new nozzles may have slight variations in actual output. Sprayer pressure gauges may not be accurate, which further adds error to the output estimate made from these charts. Replace pressure gauges that appear inaccurate or test them using commercial or homemade testers.

Measure liquid sprayer output in gallons per minute (gpm) under actual application conditions: at the correct pressure and tractor engine speed (rpm) and with all nozzles open if the sprayer has more than one. Select from one of the two methods described below, depending on the type of sprayer you are calibrating. The first method works for low-pressure sprayers, Venturi air-assisted sprayers, and small handheld units. It involves collecting a volume of water from individual nozzles over a measured time. The second method, for air carrier and high-pressure sprayers, measures the output of a sprayer over a known period.

Collection method for low-pressure and small handheld sprayers. Calibrate low-pressure sprayers by measuring the amount of spray discharged from nozzles, using clean water only. These include low-pressure boom sprayers, backpack sprayers, and controlled droplet applicators. If the sprayer has more than one nozzle, collect water from each separately, which allows you to compare each nozzle's output and points out any failure or wear. You need a stopwatch and calibrated container for making measurements. Wear chemical-resistant gloves to avoid skin contact with water that

SIDEBAR 10-2

CALCULATING THE TRAVEL SPEED OF APPLICATION EQUIPMENT

Step 1. Convert minutes and seconds into minutes by dividing the seconds (and any fraction of a second) by 60.

EXAMPLE

Your trip took 1 min and 47.5 sec.

47.5 sec ÷ 60 sec/min = 0.79 min

Add these amounts together:

1 min + 0.79 min = 1.79 min

Step 2. Get the average run time by adding the converted minutes from each run and dividing by the number of runs.

EXAMPLE

Three runs were made.

Run 1 = 1 min, 47.5 sec	=	1.79 min
Run 2 = 1 min, 39.8 sec	=	1.66 min
Run 3 = 1 min, 52.0 sec	=	1.87 min
Total	=	5.32 min

5.32 min ÷ 3 runs = 1.77 min/run average time

Step 3. Divide the measured distance by the average time. This will tell you how many feet were traveled per minute.

EXAMPLE

The measured distance is 227 feet.

227 ft ÷ 1.77 min = 128.25 ft/min

Step 4. If you wish to determine the speed in miles per hour, divide the feet per minute figure by 88 (the number of feet traveled in 1 minute at 1 mile per hour).

EXAMPLE

128.25 ft/min ÷ 88 ft/min/mph = 1.46 mph

might be contaminated with pesticide residues. Stand upwind from the nozzles to prevent fine mist or spray from contacting your face and clothing. Wear eye protection to prevent getting spray droplets in your eyes.

For low-pressure power sprayers used in agricultural applications, fill the tank at least half full with water. Start the sprayer and bring the system up to normal operating pressure. Run hydraulic agitators if they will be running during the real application. Never operate equipment outside its normal working range or you may damage the pump. If you are calibrating a PTO-driven sprayer, be sure that the tractor engine speed (rpm) is the same as that used in the ground speed calibration. If these are not the same, the pump output pressure will be different. Adjust the pressure to meet the needs of the spray situation and nozzle manufacturer's recommendations. Check the pressure by attaching a calibrated pressure gauge at either end of the boom, replacing one of the nozzles. Open the valves to all nozzles and note the pressure, make adjustments as necessary, and then remove the gauges. When all nozzles are operating at the proper pressure, collect flow for a set amount of time, typically 15 seconds to 1 minute (Fig. 10-4).

FIGURE 10-4.

To determine the output from each nozzle, collect liquid over a measured period of time. Make sure the sprayer is operating at the pressure that would be used under actual field conditions. Wear rubber gloves and eye protection, because the liquid may contain traces of pesticide.

Record the volume of liquid collected from each nozzle or orifice and the time in seconds it took to collect each amount. Use a format similar to the form in Sidebar 10-3. Determine the output in fluid ounces per second for each nozzle by dividing the volume by the number of seconds it took to collect it. Convert ounces per second into gallons per minute.

Output among nozzles will usually vary. In the example in step 1 of Sidebar 10-4, the output ranges from 0.250 gallon per minute to 0.328 gallons per minute.The variation among nozzles should not be more than 5%. The output of any nozzle should not exceed the manufacturer's rated output by more than 10%. Figure the percentage of variation as shown in the example in step 2 of Sidebar 10-4. Nozzles 3 and 5 in this example exceed these amounts and so must be replaced.

SIDEBAR 10-3

Sample Record of Nozzle Output

Nozzle	Volume (fl oz)	Time (sec)
1	12.5	23.2
2	12.0	22.5
3	15.5	24.8
4	14.5	26.1
5	19.0	27.2
6	13.0	23.9

Changing one nozzle may affect the pressure in the whole system, so if you replace a nozzle, recheck the flow rate of all the nozzles. After changing nozzles, readjust the pressure regulator to maintain the desired pressure and recalculate output as in step 1 of Sidebar 10-5.

Spray check devices are calibration aids that provide a visual representation of the spray pattern made by nozzles on horizontal spray booms. Place this portable device under a boom and collect the output from several nozzles. After collection, rotate the device from a horizontal to a vertical position. The liquid drains into a series of evenly spaced glass vials. Floats inside these vials rise to the top of the liquid. You then can see variations in liquid levels, pinpointing nozzle problems and poor nozzle height adjustment.

Measured release method for air carrier or high-pressure sprayers. Due to the air-carrying capacity and high pressures of larger sprayers, it is hard to collect the spray from the nozzles. Instead, you can find the output of the sprayer over time by measuring how much water the sprayer used.

SIDEBAR 10-4

Calculating Gallons Per Minute Output for Low-Pressure Sprayers

Step 1. Determine the gallons per minute (gpm) output of each nozzle by dividing the fluid ounces collected by the time (in seconds) and multiplying the result by 0.4688. This example uses nozzle outputs from Sidebar 10-3.

EXAMPLE

Nozzle	Output (fl oz)	÷	Time (sec)	=	Output per sec	x	0.4688	=	gpm
1	12.5	÷	23.2	=	0.539	×	0.4688	=	0.253
2	12.0	÷	22.5	=	0.533	×	0.4688	=	0.250
3	15.5	÷	24.8	=	0.625	×	0.4688	=	0.293
4	14.5	÷	26.1	=	0.556	×	0.4688	=	0.261
5	19.0	÷	27.2	=	0.699	×	0.4688	=	0.328
6	13.0	÷	23.9	=	0.544	×	0.4688	=	0.255
							Total output		1.640

Step 2. Compute the percentage of variation from the rated nozzle output. Divide the actual gallons per minute output by the rated output. Subtract 1 from this number and multiply by 100.

EXAMPLE

Nozzle	Actual gpm	÷	Rated gpm	=		–	1.00	=	Nozzle actual flow rate	×	100	=	Percent variation
1	0.253	÷	0.250	=	1.012	–	1.00	=	0.012	×	100	=	1.2
2	0.250	÷	0.250	=	1.000	–	1.00	=	0.000	×	100	=	0.0
3	0.293	÷	0.250	=	1.172	–	1.00	=	0.172	×	100	=	17.2
4	0.261	÷	0.250	=	1.044	–	1.00	=	0.044	×	100	=	4.4
5	0.328	÷	0.250	=	1.312	–	1.00	=	0.312	×	100	=	31.2
6	0.255	÷	0.250	=	1.020	–	1.00	=	0.020	×	100	=	2.0

Start by moving the sprayer to a level surface and filling the tank to the maximum measurable level that can be duplicated when refilling. A convenient method is to fill the tank with clean water to the point where it just begins to overflow. In this situation, you must keep the hose out of the water at all times (always maintain an air gap). Use low-volume, low-pressure water, such as from a garden hose, for topping off the tank. Check for leaks around tank seals and in hoses. All nozzles must be clean and operating properly or the results will be inaccurate.

FIGURE 10-5.

It is not possible to collect the sprayed liquid from some types of sprayers. To determine the amount of liquid expelled by these sprayers: (1) fill the tank to a known level; (2) run the sprayer under normal conditions for a timed period; and (3) refill the tank to its original level, measuring the amount of water used.

Stand upwind and run the sprayer at its normal operating speed and pressure. Open the valves to all nozzles, starting a stopwatch at the same time. Continue to run the sprayer for several minutes, then close the valves to all nozzles. Record the elapsed time that the nozzles were spraying (Fig. 10-5).

Use a flowmeter attached to a low-pressure filling hose or a calibrated bucket to refill the sprayer to its original level. (Using a site gauge or dipstick can result in inaccurate measurements.) Record the number of gallons of water used; this volume is the amount

SIDEBAR 10-5

RECALCULATING OUTPUT AFTER REPLACING WORN NOZZLES

Step 1. Replace worn nozzles (numbers 3 and 5 in this example) and remeasure the output of all nozzles on the boom. Recalculate the gallons per minute for each nozzle. Add these rates together to determine the total output of the sprayer.

EXAMPLE

Nozzle	Output (fl oz)	÷	Time (sec)	=	Output per sec	×	0.4688	=	gpm
1	12.5	÷	23.2	=	0.539	×	0.4688	=	0.253
2	12.0	÷	22.5	=	0.533	×	0.4688	=	0.250
3	13.3	÷	24.5	=	0.543	×	0.4688	=	0.255
4	14.5	÷	26.1	=	0.556	×	0.4688	=	0.261
5	15.2	÷	28.3	=	0.537	×	0.4688	=	0.252
6	13.0	÷	23.9	=	0.544	×	0.4688	=	0.255
							Total output	=	1.525

Step 2. Check to see that all nozzles are within 5% of the rated capacity of these nozzles.

EXAMPLE

Nozzle	Actual gpm	÷	Rated gpm	=		–	1.00	=	Nozzle actual flow rate	×	100	=	Percent variation
1	0.253	÷	0.250	=	1.012	–	1.00	=	0.012	×	100	=	1.2
2	0.250	÷	0.250	=	1.000	–	1.00	=	0.000	×	100	=	0.0
3	0.255	÷	0.250	=	1.016	–	1.00	=	0.016	×	100	=	1.6
4	0.261	÷	0.250	=	1.044	–	1.00	=	0.044	×	100	=	4.4
5	0.252	÷	0.250	=	1.008	–	1.00	=	0.008	×	100	=	0.8
6	0.255	÷	0.250	=	1.020	–	1.00	=	0.020	×	100	=	2.0

of liquid sprayed during the timed run. Repeat this process two more times to get an average of sprayer output. Determine the sprayer output in gallons per minute by using the calculations shown in Sidebar 10-6.

Swath Width. The final measurement needed to complete calibration is the width of the spray swath applied by the sprayer. Figure 10-6 shows spray swath widths for various application situations. For multiple-nozzle, boom-type sprayers, the swath is the width of the boom plus the distance between each pair of nozzles. Do not assume that all the nozzles on the boom are spaced the same distance apart. An accurate measurement of swath width must take into account the actual spacing between nozzles on the boom, so measure these distances

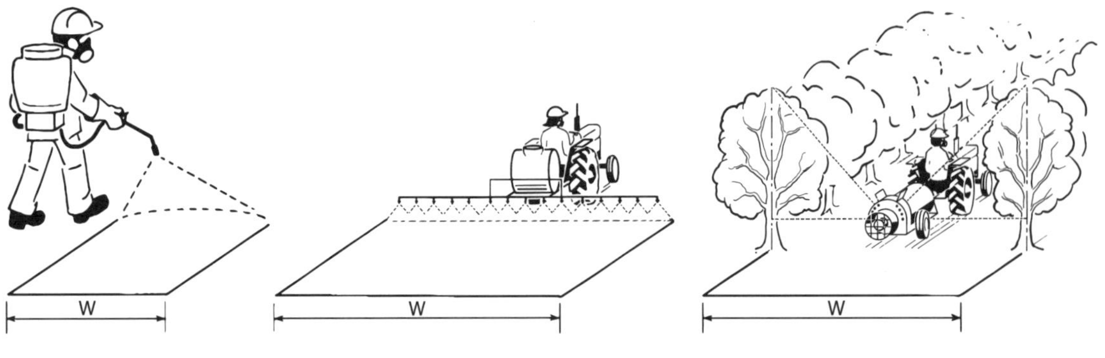

FIGURE 10-6.

A spray swath is the horizontal width being covered with spray material during a single pass. Swath width is measured differently, depending on the type of pesticide application.

SIDEBAR 10-6

Calculating Output in Gallons Per Minute for High-Pressure Sprayers

Step 1. Record the elapsed time during each trial run and the amount of liquid sprayed.

EXAMPLE

Run	Time	Volume
1	30 sec	0.48 gallons
2	30 sec	0.46 gallons
3	30 sec	0.47 gallons

Step 2. Convert time from seconds to minutes by dividing the seconds by 60.

EXAMPLE

30 sec ÷ 60 = 0.5 minute

Step 3. Divide the collected gallons for each run by the minutes to obtain gallons per minute (gpm).

EXAMPLE

Run	Gallons	÷	Minutes	=	gpm
1	0.48	÷	0.5	=	0.96
2	0.46	÷	0.5	=	0.92
3	0.47	÷	0.5	=	0.94

Step 4. Add the gallons per minute and divide this total by the number of runs (3 in this example) to get the average output in gallons per minute.

EXAMPLE

Run	Output (gpm)
1	0.96
2	0.92
3	0.94
Total	2.82
Average	0.94

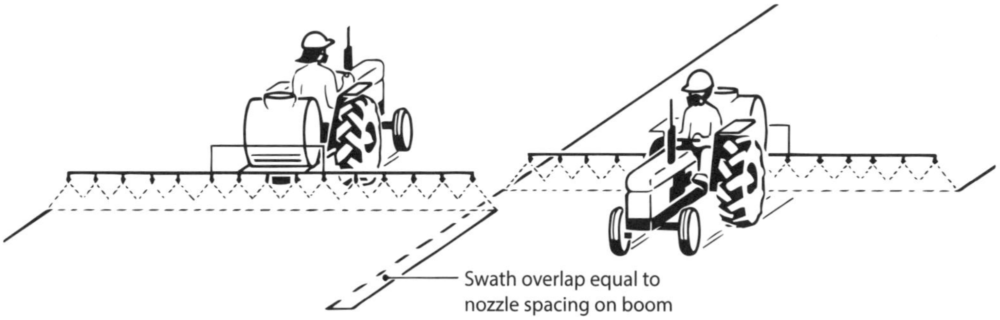

FIGURE 10-7.

Spray from adjacent swaths should overlap by the same amount as spray from nozzles on the spray boom overlaps. To account for this overlap, allow one nozzle-width spacing between swaths, as illustrated here.

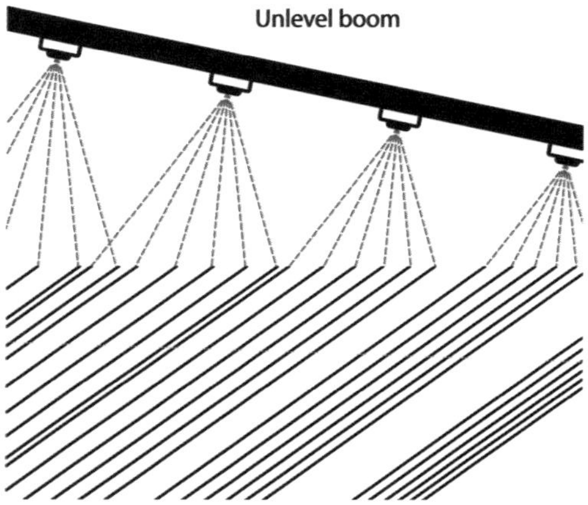

FIGURE 10-8.

An unlevel spray boom will cause an uneven pesticide application.

carefully. You can also calculate swath width by multiplying the number of nozzles by the nozzle spacing, if nozzles are evenly spaced. When making a pesticide application with a boom sprayer, overlap the spray by the same amount as the overlap of the nozzles on the boom (Fig. 10-7). Adjust the boom height so that there is sufficient overlap of spray from adjacent nozzles on the boom for even coverage. Nozzles have a specific boom height range that should be noted. Position nozzles at the exact height they would be during an actual application. Check the spray boom to make sure it is level. An unlevel boom causes uneven spray distribution and drift (Fig. 10-8). Align fan nozzles properly to give an even spray distribution (Fig. 10-9).

When applying spray as separate bands or strips, the swath width is equal to the combined width of each band. It does not include the unsprayed spaces between bands (Fig. 10-10).

When spraying crop plants on both sides of an air blast sprayer in an orchard or vineyard, the spray swath is equal to the width of the tree or vine row (Fig. 10-11). If you spray only one side of the row, the swath is one-half the width of the tree or vine spacing (Fig. 10-12). Use a tape measure to determine tree or vine row width. Take several measurements within the orchard or vineyard to check if row spacing is uniform and consistent. Average the results if you find any variation (Fig. 10-13).

Measure the swath width for herbicide strip sprays in orchards and vineyards to the center of the tree or vine row. Do not include overlap from the outside nozzle (Fig. 10-14). Unless you apply the herbicide to the entire orchard or vineyard floor, the actual sprayed area is less than the total planted area.

Some applications use an inverted U-shaped boom to apply pesticides to the tops and both sides of vines or plants in a row. Sometimes these booms cover a row on each side of the tractor. The swath width for this type of equipment is equal to the distance between opposing nozzles (Fig. 10-15).

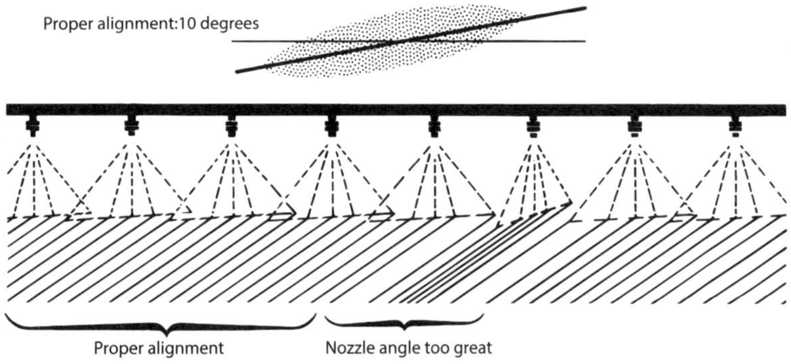

FIGURE 10-9.

The spray pattern will be uneven if nozzles are not aligned properly on the spray boom. Rotate nozzles about 10 degrees from the axis of the boom to prevent droplets from adjacent nozzles from touching, but still allow for proper overlap of the spray pattern.

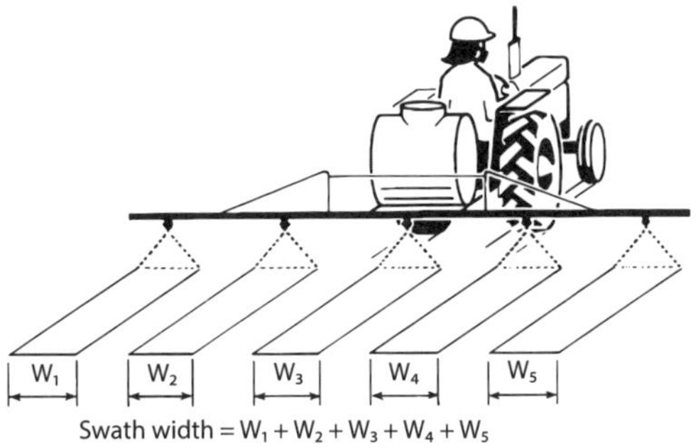

FIGURE 10-10.

Swath width from banded applications is determined by adding the widths of the individual bands.

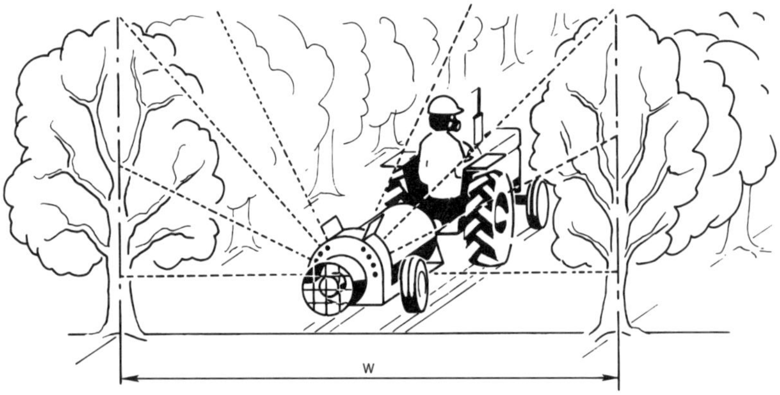

FIGURE 10-11.

In orchards or vineyards, if plants on both sides of the sprayer are being sprayed simultaneously with an air blast sprayer or high-pressure boom sprayer, the swath width is the distance between plant rows.

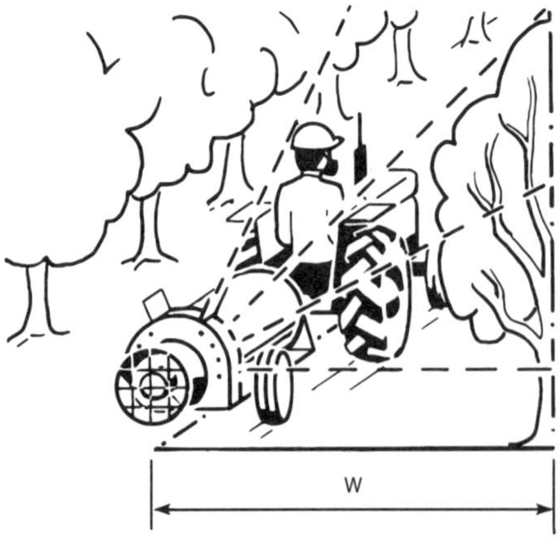

FIGURE 10-12.

When spray is emitted from only one side of an orchard or vineyard air blast sprayer, the swath width of each pass is one-half the plant row spacing.

FIGURE 10-13.

Swath width for pesticide sprays in orchards and vineyards should be measured from the center of one tree or vine row to the center of the adjacent row. Take several measurements in different locations to check for variation in plant spacing. If variation exists, average the measurements.

FIGURE 10-14.

Swath width for herbicide strip sprays in orchards and vineyards should be measured only to the center of the tree or vine row and should not include overlap.

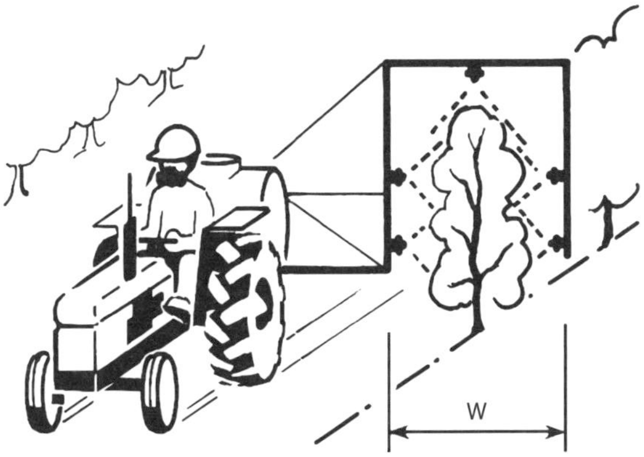

FIGURE 10-15.

Sometimes spray can be applied to both sides of a plant or vine row through a specially designed, horseshoe-shaped boom arrangement. Several plant rows can often be sprayed at the same time with these applicators. Spray swath width is the distance between opposing nozzles. If multiple rows are sprayed, the swath width is the sum of the distances.

You can inject pesticides into the soil by using special subsoil chisels spaced along a tractor-mounted toolbar. Assume that you are applying pesticides to the entire subsurface area in most soil injection applications. The swath width is equal to the number of chisels multiplied by the space between the chisels on the toolbar (Fig. 10-16). When you inject pesticides as bands, the swath width is the sum of all the band widths, similar to surface band applications.

Measure the swath width of a backpack sprayer from the spray pattern produced on the ground in a test run. Keep the nozzle at the height held during an actual application.

FIGURE 10-16.

Subsoil chisels spaced along a tractor's toolbar are used to inject pesticides into the soil. When pesticides are injected into the soil and chisels are evenly spaced across the entire toolbar, the swath width is usually considered to be the width of the toolbar.

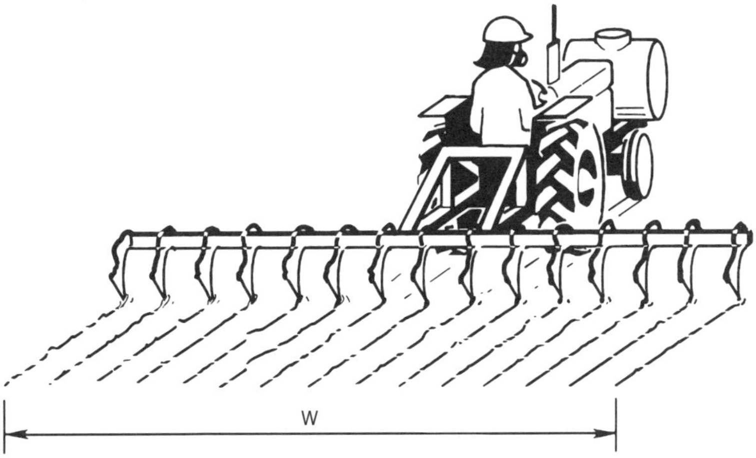

Maintain this height to prevent variation in swath width. Nozzles on these types of sprayers usually provide a uniform spray pattern. Overlap swaths only enough to ensure a uniform application pattern. Use the same method to measure swath width of controlled droplet applicators.

Describe methods used to determine how much pesticide to put into the hopper or tank for a specific application rate over the total area of the application site.

Determining the Amount of Pesticide to Use. Use tank volume, travel speed, flow rate, and swath width to calculate the total area covered with each tank of material. Knowing this value allows you to determine how much pesticide to put into the tank. Choose from two calculation methods: one for pesticides applied by the acre; the other for applications made by the square foot, such as sprays in confined areas. Sidebar 10-7 and Sidebar 10-8 describe the formulas used to make each type of calculation.

Figure 10-17 is an example of how to combine calibration formulas onto a single sheet for field use. This example shows a calibration worksheet designed for orchard sprayers. You can make similar sheets for other types of pesticide sprayers.

SIDEBAR 10-7

Calculating How Much Pesticide to Put into the Spray Tank: Pesticides Applied on a Per-Acre Basis

Step 1. Determine the area that can be treated in 1 minute. Multiply the spray swath width by the travel speed, then divide that number by 43,560 (the number of square feet in 1 acre). The result will be the acres treated per minute. In the example in Sidebar 10-2, the travel speed was calculated to be 128.25 feet per minute.

EXAMPLE

Assuming that the swath width is 12 feet, the calculation would be

(12 ft x 128.25 ft/min) ÷ 43,560 sq ft/ac = 0.0353 ac/min

Step 2. Determine the gallons of liquid being applied per acre. Divide the gallons-per-minute figure by the acres per minute

EXAMPLE

1.525 gal/min ÷ 0.0353 ac/min = 43.2 gal/ac

Step 3. Determine the number of acres that can be treated with a full tank. Divide the actual measured volume of the spray tank(s) by the number of gallons per acre being applied. The tank in this example holds 252.5 gallons when filled.

EXAMPLE

252.5 gal/tank ÷ 43.2 gal/ac = 5.84 ac/tank

Step 4. Determine how much pesticide to put in the tank. Multiply the number of acres per tank by the recommended rate per acre of pesticide; check the pesticide label for this information. If the label calls for "active ingredient," see the "Active Ingredient Calculations" section in this chapter.

EXAMPLE

Recommended rate per acre	x	Acres per tank	=	Amount of pesticide to put in tank
1.5 lb/ac	x	5.84	=	8.76 lb
3 qt/ac	x	5.84	=	17.52 qt
2 gal/ac	x	5.84	=	11.68 gal
1 pt/ac	x	5.84	=	5.84 pt

SIDEBAR 10-8

Calculating How Much Pesticide to Put into the Spray Tank: Pesticides Applied by the Square Foot

Step 1. Determine how many square feet can be treated in 1 minute. Multiply the speed as determined by the procedures in Sidebar 10-2 by the swath width. In this example, assume a single-nozzle hand-operated sprayer is being used to apply a swath width of 2.5 feet at a speed of 128.25 feet per minute.

EXAMPLE

128.25 ft/min × 2.5 ft = 320.63 sq ft/min

Step 2. Determine the volume of spray, in gallons, that will be applied to 1 square foot. Divide the gallon-per-minute output of the sprayer from Sidebar 10-4 by the number of square feet per minute. For this example, assume that the backpack unit sprays 0.05 gallon per minute.

EXAMPLE

0.05 gal/min ÷ 320.63 sq ft/min = 0.000156 gal/sq ft

Step 3. Find out how many square feet can be sprayed with one tank. Divide the number of gallons per square foot into the measured tank capacity. For this example, assume that the tank holds 3 gallons.

EXAMPLE

3 gal/tank ÷ 0.000156 gal/sq ft = 19,230 sq ft/tank

Step 4. Determine how much pesticide to put in the tank. The pesticide label will recommend the amount of pesticide to apply, normally, in the volume per square foot (or per 100 or 1,000 square feet) or per acre. If the label calls for "active ingredient," see "Active Ingredient Calculations" in this chapter.

EXAMPLE A

If the label recommends the dosage rate per 1, 100, or 1,000 square feet, multiply that rate by the square feet per tank as determined in step 3:

Label recommendation	x	Square feet per tank	=	Amount of pesticide to put in tank
3 fl oz per 1,000 sq ft	×	19,230	=	57.69 fl oz
¾* fl oz per 1,000 sq ft	×	19,230	=	14.42 fl oz
1 oz per 100 sq ft	×	19,230	=	192.3 oz

Note: *The fraction ¾ is converted to its decimal equivalent, 0.75, to complete this calculation.

EXAMPLE B

If the pesticide label recommends the dosage rate in units of pesticide per acre, convert square feet per tank (from step 3) to acres per tank by dividing it by 43,560 (the number of square feet in 1 acre):

19,230 sq ft/tank ÷ 43,560 sq ft/ac = 0.441 ac/tank

Then, multiply the label rate per acre by the number of acres per tank:

Label recommendation	x	Acres per tank	=	Amount of pesticide to put in tank
1.5 lb/ac	×	0.441	=	0.662 lb (10.6 oz)
3 qt/ac	×	0.441	=	1.323 qt (42.2 fl oz)
2 gal/ac	×	0.441	=	0.882 gal (7.1 pt)
1 pt/ac	×	0.441	=	0.441 pt (7.1 fl oz)

FIGURE 10-17.

A worksheet such as this Orchard Sprayer Calibration Worksheet can be helpful in recording and computing the figures necessary for calibration. Similar worksheets can be developed for other types of sprayers. (In this example, notice the difference between the rated output of the nozzles and the actual output. Nozzles are worn.)

ORCHARD SPRAYER CALIBRATION WORKSHEET

Grower: D. BROWN Date: 1-29-2021 Sprayer Type: AIR BLAST

CHECK:
- ☑ 1. Filter screens and strainers clean?
- ☑ 2. Tank clean and free of scale and sediment?
- ☑ 3. Pressure gauge operating?
- ☑ 4. Nozzles working properly?

Sprayer operating pressure: 100 psi

I-A. GALLONS/HOUR (Method I—using nozzle chart from manufacturer's catalog)

Nozzle Size	Number (N)		Rated Output (gallons/minute)		Minutes per Hour		Gallons per Hour
D2-25	8	×	0.25	×	60	=	120
D4-25	8	×	0.45	×	60	=	216
		×		×	60	=	
			TOTAL GALLONS PER HOUR			=	336

I-B. GALLONS/HOUR (Method 2—measurement)

1. Fill sprayer to verifiable level.
2. Run sprayer for a measured period of time (T), spraying under the same conditions as in the orchard. T = 3.53
3. Refill sprayer, measuring the amount of water used (GAL) in gallons. GAL = 20.4
4. Calculate: gallons/hour = (GAL × 60)/T TOTAL GALLONS/HOUR = 346.7

II. MILES/HOUR

1. Establish distance (D) in feet. D = 253
2. Measure elapsed time for sprayer to travel the distance. Make 3 runs and average results.
   a. First run time = 1.05 minutes.
   b. Second run time = 1.15 minutes.
   c. Third run time = 1.13 minutes.
3. Average of three runs (T) = 1.11 minutes.
4. Calculate miles per hour: MPH = (D/T)/88 MPH = 2.59

III. ACRES/HOUR

1. Measure width of tree row (W) in feet. W = 22
2. Calculate miles per acre: miles/acre = (43,560/W)/5,280 MILES/ACRE = 0.375
3. Calculate acres per hour: acres/hour = MPH/(miles/acre) ACRES/HOUR = 6.91

IV. GALLONS/ACRE

(gallons/hour)/(acres/hour) = gallons/acre GALLONS/ACRE = 50.17

V. ACRES/TANK

Tank size = 500 gallons/tank

(gallons/tank)/(gallons/acre) = acres/tank ACRES/TANK = 9.97

VI. AMOUNT OF PESTICIDE/TANK

Recommended amount of pesticide/acre = 2.5 lb.

(pesticide/acre) × (acres/tank) = pesticide/tank PESTICIDE/TANK = 24.9 lb.

VII. CALIBRATION CHECK

1. Tree spacing (S) = 22 × 22 feet S = 484
2. Trees per acre (T) = 43,560/S T = 90
3. Count the actual number of trees sprayed (N) with one tank: N = 918
4. Actual acres sprayed = N/T ACTUAL ACRES = 10.2
5. Calculated acres per tank (from "V" above) CALCULATED ACRES/TANK = 9.97
6. Percent accuracy = calculated acres/actual acres x 100 ACCURACY = 97.7%

Active Ingredient: Chlorothalonil (tetrachloroispthalonitrile)	40.4%
Inert Ingredients:	59.6%
Total	100.0%

Keep Out of Reach of Children
WARNING – AVISO
Si usted no entiende la etiqueta, busque a alguien
(If you do not understand the label, find someone to explain it to you in detail.) See side panel for additional precautionary statements.

Changing Sprayer Output

Describe the best way to change the output of various pesticide application equipment and the consequences of each change.

Once you calibrate a sprayer, you have determined its output rate for a specific speed, nozzle, and pressure. There may be times when you need to change this output rate. These include

- variations in foliage
- different plant spacing
- special needs of the treatment area
- variations in travel speeds
- compensations for nozzle or pump wear

You can make several changes, either alone or in combination, to effectively increase or decrease sprayer output within a limited range. You should also take steps to assess whether or not the pesticide you will apply is reaching its target. There are several methods you can use to check whether or not the spray has penetrated into dense foliage or determine if the spray reached its target. These are covered in Chapter 11, "Follow-up Monitoring."

Changing Speed. The easiest way to adjust the volume of spray being applied is to change the travel speed of the sprayer. A slower speed increases the application rate, while a faster speed reduces the application rate. You may need these adjustments when swath width changes slightly. This would be the case in orchards or vineyards where plant spacing differs from block to block (Fig. 10-18). Changing the travel speed means you will not have to change the concentration of chemical in the spray tank. However, there are limits to the amount of speed change you can make. Operating application equipment too quickly is a common error and results in poor coverage. Operating it too slowly results in runoff, waste, and an increase in application time and cost. To determine how much to increase or decrease your speed, put in the new swath width and rework the calculations shown in Sidebar 10-7 or Sidebar 10-8. However, if you are making an application to a vineyard or orchard with a sprayer that uses a fan, speed may not be a good way to change output. In this case, when ground speed is increased, the fan will not have enough time to blow the pesticide into the canopy fully. So use caution when changing speed to modify output.

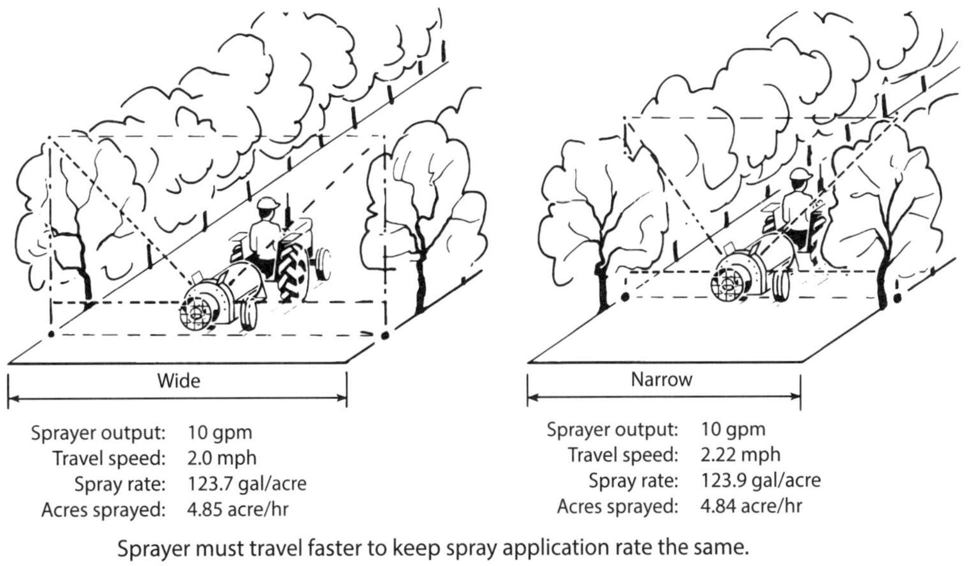

FIGURE 10-18.

Changes in row spacing in an orchard or vineyard affect the amount of spray being applied per acre. Increasing or decreasing ground speed can accommodate for the difference in spacing so the correct amount of pesticide per acre will be applied. Variations in the size of trees or vines may also influence the rate of application.

Changing Nozzle Size. The most effective way to change the output volume of a sprayer is to install nozzles of a different size. Larger nozzles increase volume, while smaller ones reduce volume. Changing nozzles can change the pressure of the system, which may require an adjustment of the pressure regulator. Adjust the output volume of disc-core nozzles by changing either the disc or the core. Sometimes you need to replace both. Be aware that changes in either the core or disc also change the droplet size and spray pattern. Use tables included in nozzle manufacturers' catalogs as a guide for estimating the output of different combinations. Whenever you change any nozzles, recalibrate the sprayer and recalculate its new total output.

Changing Output Pressure. Adjusting the pressure regulator to increase or decrease output pressure changes the spray volume slightly: increasing pressure increases output, while decreasing pressure lowers it. However, to double the output volume you must increase the pressure by a factor of four. This is usually beyond the working pressure range of the sprayer pump. Whenever pressure in the system changes, measure the nozzle output again (see Sidebar 10-4). Then, rework the calibration calculations. Increasing pressure breaks the spray up into smaller droplets, increasing the likelihood of spray drift. Lowering pressure too much reduces the effectiveness of nozzles by reducing their ability to form appropriately sized droplets.

Calibrating Dry Applicators

Describe how to calibrate dry applicators.

The methods for calibrating dry applicators are similar in many ways to those used for liquids. Granules vary in size and shape from one pesticide to the next, influencing their flow rate from the applicator hopper. You should calibrate granule applicators for each type of granular pesticide you apply. Also, recalibrate this equipment each time weather or field conditions change, especially if humidity increases.

Before starting to calibrate a dry applicator, be sure that it is clean and all parts are working properly. Most equipment requires periodic lubrication. Calibrating granule applicators involves using actual pesticides, so wear the label-prescribed PPE. Some formulations are dusty and may require respiratory protection. Always wear chemical-resistant gloves to prevent contact with residues on the equipment. You must measure three variables when calibrating a dry applicator:

- travel speed
- output rate
- swath width

Travel Speed. Determine travel speed in feet per minute in the same manner as you would for liquid applicators. Follow the instructions given in Sidebar 10-2.

Output Rate. To determine the output rate, fill the hopper or hoppers with the granular pesticide. Most granule applicator hoppers have ports with adjustable openings for granules to pass through. Refer to the manufacturer's instructions to determine the approximate opening for the application rate and speed you need. Once you set the approximate opening, use one of the following three methods to determine the actual output rate.

1. *Measure the amount of granules applied to a known area.* The easiest way to calibrate a granule applicator is to collect and weigh the granules applied to a known area. Use this method when working with broadcast applicators. Spread out a plastic tarp of known size on the ground. Then operate the broadcast applicator at a known speed across the tarp (Fig. 10-19). Place the granules collected by the tarp into a container and weigh them. Use the calculations shown in the example in Sidebar 10-9 to figure the amount of granular pesticide applied per acre or other unit of area.

2. *Collect a measured amount of granules over a known period of time.* Collecting and weighing measured amounts of granules is similar to calibrating a liquid boom sprayer with multiple nozzles. Use this method for granule applicators with multiple ports. While operating the applicator at a normal speed, collect granules from one port at a time. Record the time needed to

FIGURE 10-19.

To determine the area of granules being applied, measure the swath width across a plastic tarp and multiply this by the length of the tarp.

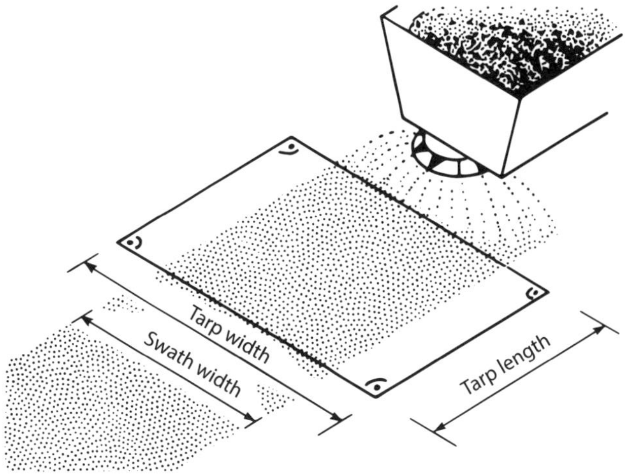

Area measurement: swath width × tarp length

SIDEBAR 10-9

Calculating Granule Output Rate by Measuring the Quantity Applied to a Known Area

Step 1. Spread a large plastic tarp on the ground. Make sure the tarp is wide enough to contain all the granules that will be distributed by the applicator and is at least 10 feet long. This example uses a tarp 15 feet wide and 10 feet long.

Step 2. Fill the hopper or hoppers of the applicator, adjust the output ports to the correct opening according to the label, and move the applicator across the entire length of the tarp at an even pace while broadcasting granules. Note the travel speed.

Step 3. To determine the swath area, measure the swath width of the granules on the tarp and multiply it by the length of the tarp (the distance traveled). In this example, the swath width is 12 feet.

EXAMPLE

Swath area = 10 x 12 = 120 sq ft

Step 4.

1. Transfer the granules on the tarp to a container and weigh them. In this example, the granules on the tarp weighed 4 ounces.
2. Convert the output in the swath area to a per-acre application rate by converting the weight of the granules to pounds and multiplying by 43,560 (the number of square feet in 1 acre) and dividing by the swath area.

EXAMPLE

4 oz ÷ 16 oz (1 lb) = 0.25 lb

0.25 lb x 43,560 sq ft ÷ 120 sq ft = 90.75 lb/ac

In this example, the applicator is broadcasting 90.75 pounds of granules per acre. The label of this product calls for an application rate of 80 pounds per acre. To lower the application rate to the listed rate, close the port some or increase the travel speed. Once an adjustment has been made, repeat the calibration procedure to ensure your applicator is calibrated to apply the exact rate required by the label.

SIDEBAR 10-10

Calculating Granule Output Rate by Collecting a Measured Amount over a Known Period of Time

Step 1. Adjust the hopper opening according to the manufacturer's instructions suggested for your required application rate. If no information is available, begin with an intermediate setting.

Step 2. Operate the equipment at the speed of an actual application. Collect granules in a clean container, such as a pan or bag, before they drop to the ground. Use a stopwatch to determine the time required to collect each volume. If granules are dispersed through more than one opening, collect and time the output from each. Because some units drop granules onto a spinning disc for dispersal, it may be necessary to disable the disc by disconnecting the drive chain or belt to prevent granule loss during collection. For smaller units, collect the discharge in a bag placed over the outlet. Be sure granules are moved away from the port quickly enough to prevent clogging.

Step 3. Weigh the output from each port separately to detect any variability; if necessary, adjust ports to equalize flow rates. Collections should be weighed in ounces.

Step 4. Determine the output in pounds per hour. Divide each weight by the collection time and multiply by 0.0625 (the number obtained by dividing 1 minute by 16 ounces per pound; this number will convert ounces per minute into pounds per minute).

EXAMPLE

The following is an example of an output collected from a granule applicator with six ports, although the same calculations would apply if only one port were used. Hopper openings were adjusted following the manufacturer's instructions for an application of 200 pounds per acre:

Port	Output (oz)	Time (min)
1	29.5	0.25
2	33.0	0.28
3	31.5	0.26
4	29.0	0.25
5	33.0	0.27
6	30.0	0.26

Port	Output (oz)	÷	Time (min)	=	Output (oz/min)	×	0.0625	=	Output (lb/min)
1	29.5	÷	0.25	=	118.0	×	0.0625	=	7.375
2	33.0	÷	0.28	=	117.9	×	0.0625	=	7.369
3	31.5	÷	0.26	=	121.2	×	0.0625	=	7.575
4	29.0	÷	0.25	=	116.0	×	0.0625	=	7.250
5	33.0	÷	0.27	=	122.2	×	0.0625	=	7.638
6	30.0	÷	0.26	=	115.4	×	0.0625	=	7.213
								Total output per hour	44.420

Step 5. Determine the total output in pounds per minute by adding the individual outputs of each port. In this example the total output is 44.42 pounds per minute.

Step 6. Use the technique shown in Sidebar 10-12 to calculate the rate per acre or other unit of area.

collect each sample. Weigh samples separately, then use the calculations shown in Sidebar 10-10 to find the output rate. This method can be more efficient than others.

3. *Refill the hopper after a measured period of time.* Use this method with hand-operated equipment or when applying small amounts. It also works best when you have several applicators on a boom. Fill the hopper or hoppers to a known level, and operate the equipment for a measured time. When finished, weigh the amount of granules needed to refill the hoppers to their original levels. Use the calculations shown in the example in Sidebar 10-11 to compute the output rate. Settling of granules in the hoppers may cause this method to be less accurate than the first two methods described above.

Swath Width. To measure the swath width, operate the equipment under actual field conditions. Whenever possible, place cans, trays, or other containers at even intervals across the application swath to collect granules. Weigh the granules collected in each container separately to determine the distribution pattern. You can operate some spreaders over a strip of black cloth or plastic. This gives you a rapid visual assessment of granule distribution and swath width. Applicators that apply bands or inject granules into the soil do not have devices to disperse granules from side to side. Determine swath width by adding the widths of individual bands.

SIDEBAR 10-11

Calculating the Rate of Output by Refilling the Hopper after a Measured Period of Time

Step 1. Fill the hopper or hoppers to a known level with granules.

Step 2. Operate the equipment for a measured period of time at a known speed.

Step 3. Weigh the amount of granules required to refill the hopper or hoppers to their original level. If multiple hoppers are being used, be sure each is applying approximately the same amount of granules. If a significant variation exists, adjust the ports and repeat steps 1 through 3.

EXAMPLE

In this example, six applicators are used together on a boom. They have been adjusted so that they all apply approximately the same amount of granules.

Hopper	Operating time (min)	Weight of granules (lb)
1	2.5	6.2
2	2.5	6.1
3	2.5	6.1
4	2.5	6.3
5	2.5	6.1
6	2.5	5.9
	Total	36.7

Step 4. Convert the output to pounds per minute by dividing the total weight from all hoppers by the time they were operated.

EXAMPLE

$$36.7 \text{ lb} \div 2.5 \text{ min} = 14.68 \text{ lb/min}$$

Step 5. Use the technique shown in Sidebar 10-12 to calculate the rate per acre or other unit of area.

Application Rate. Use the example in Sidebar 10-12 to calculate the actual rate of granules being applied per acre or other unit of area. If your calculations do not correspond to the labeled rate, adjust the equipment and repeat the calibration procedure. Motorized and hand-operated applicators apply granules at a fixed output, independent of ground speed. When ground speed increases, you apply fewer granules per unit of area. When ground speed decreases, you apply more material. With this type of equipment, you can adjust the application rate by adjusting the size of the port opening and by changing the speed of travel.

The output of ground wheel-driven granule applicators varies according to the ground speed. If ground speed increases, the applicator runs faster and the output rate is greater. When the ground speed slows down, output decreases because the applicator runs slower. The result of this automatic change in output is that the equipment applies nearly the same amount of material per acre or other unit of area no matter what speed it travels. The equipment has minimum and maximum operating speeds determined by the manufacturer, however. You change the application rate by increasing or decreasing the size of the port openings. In some units, you also change drive gears or sprockets to change the speed of the metering mechanism.

Calculation for Active Ingredient, Percentage Solutions, and Parts per Million Dilutions

Before adding pesticide to the spray tank, read and understand the dilution instructions on the label.

Active Ingredient Calculations

Describe how to determine the correct amount of pesticide needed for a particular application, including diluting a pesticide correctly.

Pesticides are seldom available in their pure state. Manufacturers formulate them into a pest control product by combining them with adjuvants and other ingredients. Therefore, only a portion of any formulated product, whether dry or liquid, is made up of the active ingredient (a.i.). Some pesticide use guidelines, including those published by the University of California, call for a specific a.i. if there are several formulations available. Because different manufacturers sell different formulations, using a.i. calculations allows you to apply the same amount of actual pesticide to a unit of area no matter what formulation you use.

SIDEBAR 10-12

Calculating Application Rate Per Acre or Other Unit of Area

Step 1. Determine the acres per minute being treated by dividing the swath width by 43,560 (the number of square feet in an acre) and multiplying the result by the speed of travel. In this example, the swath width is 30 feet and the application speed is 352 feet per minute (4 miles per hour).

EXAMPLE

$$(30 \text{ ft} \div 43{,}560 \text{ sq ft/ac}) \times 352 \text{ ft/min} = 0.242 \text{ ac/min}$$

Step 2. Determine the pounds of formulated pesticide being applied per acre by dividing the output rate of the granule applicator (as computed from the calculations in Sidebar 10-9, 10-10, or 10-11) by the acres per minute calculated in step 1. This example uses 44.42 pounds per minute as the output rate.

EXAMPLE

$$44.42 \text{ lb/min} \div 0.242 \text{ ac/min} = 183.6 \text{ lb/ac}$$

Be able to calculate the active ingredient concentration of pesticides using formulas.

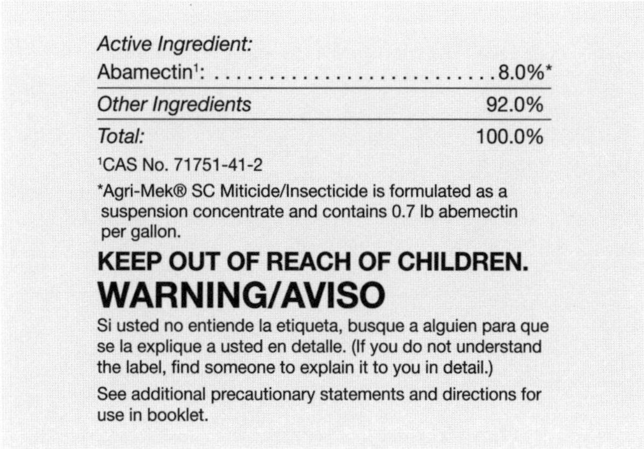

Active Ingredient:

Abamectin[1]: 8.0%*

Other Ingredients	92.0%
Total:	100.0%

[1]CAS No. 71751-41-2

*Agri-Mek® SC Miticide/Insecticide is formulated as a suspension concentrate and contains 0.7 lb abemectin per gallon.

KEEP OUT OF REACH OF CHILDREN.

WARNING/AVISO

Si usted no entiende la etiqueta, busque a alguien para que se la explique a usted en detalle. (If you do not understand the label, find someone to explain it to you in detail.)

See additional precautionary statements and directions for use in booklet.

FIGURE 10-20.

To determine the percentage of active ingredient in a pesticide formulation, check the pesticide label. Labels for all formulation types list the active ingredient as the total percentage of the weight. Liquid formulations, like this one, also list the number of pounds per gallon of formulation.

Manufacturers list the percentage of a.i. on product labels. Labels of pesticides give the percentage by weight of a.i. (Fig. 10-20). The labels of liquid pesticides also tell how many pounds of a.i. are in 1 gallon of formulation. Use the calculations in Sidebar 10-13 to make a.i. calculations with liquid formulations. Use Sidebar 10-14 for dry (powder) formulations and Sidebar 10-15 for granular formulations.

Percentage Solutions

Sometimes labels require that the pesticide be mixed as a percentage solution. You mix the product to get a known concentration regardless of the sprayer output rate. Mix percentage solutions on a weight-to-weight (w/w) basis, meaning pounds of a.i. per pound of water. Sidebar 10-16 provides an example of calculating a percentage solution with liquid formulations. Sidebar 10-17 shows the calculations for dry formulations.

SIDEBAR 10-13

Calculating Liquid Formulations

Assume that a sprayer has been calibrated and found to spray 7.5 acres per tank. You have a recommendation to apply 1.5 pounds of a.i. of chlorothalonil per acre to control rust on snap beans and have been supplied with a liquid formulation containing 4.17 pounds a.i. per gallon. How much chlorothalonil should you put in the tank?

Step 1. Determine the number of gallons of liquid needed per acre by dividing 1 gallon by the pounds of a.i. per gallon and multiplying that by the pounds a.i. per acre.

EXAMPLE

(1 gal ÷ 4.17 lb/gal a.i.) x 1.5 lb a.i./ac = 0.360 gal/ac

Step 2. Multiply the known acre capacity of the tank by the gallons per acre.

EXAMPLE

7.5 ac/tank x 0.360 gal/ac = 2.7 gal/tank

This is the number of gallons of formulated chlorothalonil that should be put into the tank for spraying 7.5 acres of crop.

SIDEBAR 10-14

Calculating Powder Formulations

The calibrated sprayer you are using covers 7.5 acres per tank, and you have a recommendation to apply 1.5 pounds a.i. of chlorothalonil per acre for control of rust on snap beans. You are provided with a wettable powder formulation that, according to the label, contains 75% chlorothalonil. How much chlorothalonil should you put into the tank?

Step 1. Convert the percentage of a.i. to a decimal by dividing by 100 (or simply move the decimal point two places to the left).

EXAMPLE

75% = 0.75 lb a.i./lb formulation

Step 2. Divide the recommended amount of a.i. by the amount of a.i. in the formulation.

EXAMPLE

1.5 lb a.i./ac ÷ 0.75 lb a.i./lb formulation = 2 lb formulation/ac

Step 3. Multiply the pounds of formulation per acre by the number of acres per tank to find out how much material to put into the tank.

EXAMPLE

2 lb formulation/ac × 7.5 ac/tank = 15 lb/tank

SIDEBAR 10-15

Calculating Granular Formulations

You are given a recommendation to apply 0.50 lb a.i. of ethoprop per 1,000 square feet of turf for control of nematodes. You are provided with a granular formulation containing 10% active ingredient (0.1 pound of a.i. per pound of formulation). To what rate should you calibrate the granule applicator?

Step 1. Convert the percent a.i. to a decimal and divide this into the recommended application rate.

EXAMPLE

0.5 lb a.i. per 1,000 sq ft ÷ 0.1 lb a.i. per lb formulation =
5 lb formulation

Step 2. Calibrate the granule applicator so that it applies 5 pounds of formulated ethoprop per 1,000 square feet.

SIDEBAR 10-16

Calculating a Percentage Solution: Liquid Formulations

To prepare a percentage solution using liquid formulations, you need to know

- the volume of the spray tank
- the weight of a.i. per gallon of formulation
- the weight of a gallon of water

The weight of water is a constant, approximately 8.34 pounds per gallon. Assume you have measured the volume of the spray tank and find that it holds 264.5 gallons of water. You are given a recommendation to apply a 1% solution of glyphosate for control of aquatic weeds using a high-pressure sprayer with a handheld spray nozzle. The formulation of glyphosate that you are to use contains 5.4 pounds of a.i. per gallon.

Step 1. Find the total weight of the liquid in the filled tank by multiplying the volume of the tank (264.5 gallons) by the weight of water (8.34 pounds per gallon).

EXAMPLE

264.5 gal × 8.34 lb/gal = 2,205.93 lb

Step 2. Multiply this weight by 0.01 (1%) to determine the weight of a.i. required to mix a 1% solution.

EXAMPLE

2,205.93 × 0.01 = 22.06 lb

Step 3. Divide the required weight of a.i. by the weight of a.i. in the formulation. The result is the number of gallons of liquid formulation that should be added to 264.5 gallons of water to achieve a 1% solution.

EXAMPLE

22.06 lb a.i. ÷ 5.4 lb a.i./gal = 4.1 gal formulation

In this example, one tank of liquid should contain 4.1 gallons of glyphosate formulation. The total volume of water combined with the glyphosate formulation should equal 264.5 gallons, the capacity of the tank. You would therefore use 260.4 gallons of water and 4.1 gallons of formulated glyphosate.

Note: These calculations give a close approximation of the amount of liquid formulation to add to the tank to achieve a known percentage solution. The mathematics for a more exact figure are more complex and unnecessary for this type of work.

SIDEBAR 10-17

CALCULATING A PERCENTAGE SOLUTION: DRY FORMULATIONS

The calculations for percentage solutions using dry formulations are similar to the calculations for liquid formulations. First, from the label, determine the percentage of a.i. in the dry formulation. Assume for this example that it is 75% a.i.; 1 pound of dry formulation would contain 0.75 pound of pesticide a.i. You need to mix a 1% spray solution of this formulation in a 264.5-gallon tank.

Step 1. Find the total weight of the liquid in the filled tank by multiplying the volume of the tank by the weight of water per gallon.

EXAMPLE

$$264.5 \text{ gal} \times 8.34 \text{ lb/gal} = 2{,}205.93 \text{ lb}$$

Step 2. Multiply this weight by 0.01 (1%) to determine the weight of a.i. required to mix a 1% solution.

EXAMPLE

$$2{,}205.93 \times 0.01 = 22.06 \text{ lb}$$

Step 3. Divide the weight of a.i. by the decimal equivalent of the percentage of a.i. in the formulation. The result is the number of pounds of formulation that should be added to 264.5 gallons of water to achieve a 1% solution.

EXAMPLE

$$22.06 \text{ lb} \div 0.75 = 29.41 \text{ lb formulation}$$

Step 4. Add 29.41 pounds of wettable powder to 264.5 gallons of water to achieve a 1% solution.

TABLE 10-1:

Parts per million (ppm)

Parts per million (ppm)	Decimal solution	Percentage
1 ppm	0.000001	0.0001%
10 ppm	0.00001	0.001%
100 ppm	0.0001	0.01%
1,000 ppm	0.001	0.1%
10,000 ppm	0.01	1.0%
100,000 ppm	0.1	10.0%
1,000,000 ppm	1.0	100.0%

Explain how system controllers can impact the calibration of equipment and calculations necessary to apply pesticides effectively.

Explain the importance of properly calibrating sensors that are part of a system controller.

PARTS PER MILLION DILUTIONS

You must mix certain pesticides in parts per million (ppm) concentrations. These are the same as percentage solutions. For example, a 100 ppm dilution is equal to a 0.01% solution (Table 10-1). The ppm designation represents the parts of a.i. of pesticide per million parts of water. Parts per million dilutions are a common way of measuring very diluted concentrations of pesticides. When calculating parts per million, use the formulas in Sidebar 10-18 if you are mixing dry formulations with water. For liquid formulations, use the formulas in Sidebar 10-19.

Using System Monitors and Controllers

System monitors and controllers are becoming more popular in achieving accurate application, but they do not eliminate the need for sprayer inspection and calibration. Monitors measure operating conditions such as travel speed, pressure, and/or flow rate and can alert you to unexpected changes in application rates. Rate (or spray) controllers are monitors with the added capability of automatic rate control. The controller receives the actual application rate from the monitors and compares it with the desired rate. If an error exists, the pressure is regulated to adjust the spray volume. However, nozzles can operate only within a limited range of pressure without either distorting the spray angle or creating off-target drift, so you must be observant during applications to ensure that spray remains even and on target.

Do not assume that the monitors are foolproof. Consult the manufacturer's operator's manual to properly calibrate and adjust the sensors. Monitors that give travel speed, spray volume, and so on are usually adequate for most sprayer situations. The newer monitors keep track of which booms are being used and the areas they cover so that the calculated area sprayed is very accurate.

SIDEBAR 10-18

Calculating a Parts per Million Dilution: Dry Formulations

You are given a recommendation for a 100 ppm concentration of oxytetracycline to be mixed in a 500-gallon tank for control of fire blight on pear trees. The formulation you have is a wettable powder containing 17% a.i.

Step 1. Find the total weight of the liquid in the filled tank by multiplying the volume of the tank by the weight of water per gallon.

EXAMPLE

500 gal × 8.34 lb/gal = 4,170 lb/tank

Step 2. Determine how many pounds of a.i. are required for 1 pound of spray solution.

EXAMPLE

100 ppm = 100 parts a.i. ÷ 1,000,000 parts solution = 0.0001

Step 3. Determine how many pounds of a.i. are required for a tank of solution, using the weight of the liquid in the tank.

EXAMPLE

4,170 lb/tank × 0.0001 lb a.i. = 0.417 lb a.i.

Step 4. Divide the weight of a.i. by the decimal equivalent of the percentage of a.i. in the formulation. The result is the number of pounds of formulation that should be added to 500 gallons of water to achieve a 100 ppm solution.

EXAMPLE

0.417 lb a.i. ÷ 0.17 lb a.i./lb formulation = 2.45 lb formulation

SIDEBAR 10-19

Calculating a Parts per Million Dilution: Liquid Formulations

Assume that a pesticide contains 5.4 pounds of a.i. in 1 gallon of formulation. You are required to prepare a 100 ppm concentration in a 500-gallon tank.

Step 1. Find the total weight of the liquid in the filled tank by multiplying the volume of the tank by the weight of water per gallon.

EXAMPLE

500 gal/tank × 8.34 lb/gal = 4,170 lb/tank

Step 2. Determine how many pounds of a.i. are required for 1 pound of spray solution.

EXAMPLE

100 parts a.i. ÷ 1,000,000 parts solution = 0.0001

Step 3. Determine how many pounds of a.i. are required for a tank of solution using the weight of the liquid in the tank.

EXAMPLE

4,170 lb/tank × 0.0001 lb a.i. = 0.417 lb a.i./tank

Step 4. Divide the required weight of a.i. by the pounds of a.i. per gallon to determine how many gallons of formulation are required. Since this will probably be a small number, convert to ounces by multiplying the result by the number of ounces per gallon (128).

EXAMPLE

0.417 lb a.i./tank ÷ 5.4 lb a.i./gal = 0.0772 gal/tank

0.0772 gal/tank × 128 fl oz/gal = 9.88 fl oz/tank

Adding 9.88 fluid ounces of this formulated pesticide to 500 gallons of water will result in a 100 ppm solution.

Chapter 10 Review Questions

1. Calibration is defined as what you must do before an application to be sure that you ____________________.

- ☐ a. avoid hazards and sensitive areas during pesticide applications
- ☐ b. select the most effective pesticide to apply in a situation
- ☐ c. apply the correct amount of pesticide to the treatment area

2. Which tools do you need to accurately and professionally calibrate pesticide application equipment? Select all that apply.

- ☐ a. droplet size chart
- ☐ b. stopwatch
- ☐ c. measuring tape
- ☐ d. wooden toothpick
- ☐ e. calibrated container
- ☐ f. magnifying glass
- ☐ g. flowmeter
- ☐ h. flagging tape

3. It takes your equipment 3 minutes to travel 264 feet. How fast, in miles per hour (mph), is the equipment travelling?

- ☐ a. 1 mph
- ☐ b. 2 mph
- ☐ c. 3 mph

4. By measuring the output of each nozzle on the spray boom, you discover that the sprayer output is 256 ounces in 30 seconds. What is the output of the sprayer in gallons per minute (gpm)?

- ☐ a. 2 gpm
- ☐ b. 3 gpm
- ☐ c. 4 gpm

5. Your calibrated sprayer with a 300-gallon tank will cover 4.2 acres. You plan to apply an herbicide at a label rate of 1.5 pounds per acre. How much of this herbicide will you put into the spray tank?

- ☐ a. 4.2 pounds
- ☐ b. 6.3 pounds
- ☐ c. 8.5 pounds

6. Which three variables must be measured when calibrating dry application equipment?

- ☐ a. output rate, hopper size, and formulation type
- ☐ b. formulation type, travel speed, and output rate
- ☐ c. swath width, output rate, and travel speed

CHAPTER 11
USING PESTICIDES EFFECTIVELY

Predicting Pest Problems 154
Making Pesticide Use Decisions 154
Choosing the Right Pesticide 155
Applying Pesticides Effectively 158
Mixing Pesticides 159
Pesticide Resistance 161
Preventing Offsite Movement of Pesticides 162
Chapter 11 Review Questions 170

Knowledge Expectations

1. Explain how monitoring and economic injury data influence pesticide use decisions.
2. Describe how to choose the most appropriate pesticide for a particular application so that the application is maximally effective and hazards are reduced.
3. Explain how a GPS unit can impact the effectiveness of pesticide applications.
4. Explain how to determine whether two or more pesticides will be compatible for tank mixing.
5. Describe mixing procedures for
   a. a single pesticide
   b. two or more pesticides (tank mix)
6. List the factors that contribute to pesticide resistance.
7. Describe the different types of offsite movement, including factors that can affect the occurrence of each type.
8. Describe how to evaluate spray coverage and adjust application variables to change coverage as needed.

Predicting Pest Problems

Explain how monitoring and economic injury data influence pesticide use decisions.

Early detection through monitoring lets you plan a program for following pest development and activity, which will help you predict if or when treatment is necessary. It also helps if you review the pest history of the farm where you work. Then you will know what pests to expect at different times of the year. If this information is not available, try to get pest history information from a similar location nearby.

When monitoring for pests, look for conditions that favor pest buildup. For example, some pest insects overwinter in crop residues or field borders. If you see these pests in such areas, there is a strong likelihood they will move into the crop. Weeds that produce seeds provide a seed reservoir for the following year. If this is the case, anticipate large populations of these weeds in following seasons. This type of monitoring will help you figure out when economic injury (or treatment) thresholds have been, or might be, reached. Predicting when a site might experience expensive pest damage helps you apply pesticides only when that application makes economic sense.

Recognizing Key Life History Information. When you monitor pest populations, you learn to recognize their life cycles and stages of development. This information is useful when planning a management program, because success depends on using the right control method at the right time. Some of the things you might learn include the pest's

- preferred habitat
- food and moisture preferences
- time of greatest activity
- seasonal occurrences and life stages

Monitoring Weather. Weather greatly influences development of plants and their pests. Wetness from rain, fog, or irrigation is a primary factor that favors most diseases. Temperature is one of several factors that control plant growth, and it determines the rate at which invertebrate organisms develop. For instance, insects complete a generation in less time when temperatures are higher than when weather is cool. The link between the weather data and the rate of arthropod pest development is covered on the UC IPM website at ipm.ucanr.edu/WEATHER/index.html.

You can find up-to-date weather information from many sources. The California Department of Water Resources CIMIS (California Irrigation Management Information System) program monitors weather variables in many locations throughout the state and reports them online on the California Department of Water Resources website. Also, other organizations, including newspapers, radio stations, and universities, redistribute CIMIS weather data. National Weather Service broadcasts local and regional weather observations and forecasts on National Oceanic and Atmospheric Administration (NOAA) Weather Radio (VHF channels 162.42, 162.50, or 162.55 MHz, depending on location).

Making Pesticide Use Decisions

Describe how to choose the most appropriate pesticide for a particular application so that the application is maximally effective and hazards are reduced.

How do you decide when to use a pesticide and what pesticide to use? Answer the following questions before making any pesticide use decision:

- What pest is affecting the crop, and how advanced is the infestation?
- What is the growth stage of the pest? The crop?
- What are the physical and weather conditions at the application site?
- What hazards are associated with the pesticide formulations available?
- What effect will the pesticide have on beneficial organisms? Secondary pests?
- What type of application equipment is available to you?
- What will it cost to purchase and apply the identified material, and is that cost less than that of other control methods you might consider?

Choosing the Right Pesticide

Choosing the right pesticide or combination of pesticides can be hard. Often you can choose from several pesticides to control a pest in a particular situation. To get information about pesticides for specific uses, consult

- sources of labels such as Internet-based label databases and manufacturers' websites
- University of California farm advisors and county agricultural commissioners
- licensed pest control advisers
- pesticide chemical handbooks
- University of California print and online publications, treatment guides, and Pest Management Guidelines (Fig. 11-1)

The University of California Statewide Integrated Pest Management Program's (UC IPM's) agricultural Pest Management Guidelines, including pesticide recommendations, are accessible through the Internet at ipm.ucanr.edu (Fig. 11-2). Up-to-date pesticide use guidelines, pesticide toxicology information, and other pest management methods are accessible through this website.

FIGURE 11-1.

Pest management manuals are published by the University of California for many different types of crops. These are useful for selecting the proper chemical and other control methods.

FIGURE 11-2.

The University of California Statewide Integrated Pest Management Program maintains a listing of pest management guidelines that can be accessed through the Internet.

Factors to Consider When Selecting a Pesticide

Pesticide Selectivity

Selectivity refers to the range of organisms affected by a pesticide. A broad-spectrum (or nonselective) pesticide kills a wide range of pests as well as nontarget species. A selective pesticide controls a smaller group of more closely related organisms, often leaving beneficials and nontarget organisms unharmed.

Toxicity of the Pesticide to be Used

As a rule, and if you have a choice, select pesticides with the signal word that indicates the lowest level of hazard. DANGER indicates the highest level of hazard, and WARNING, the next highest level. CAUTION pesticides are preferable and will usually be safer for you to work with. Often, they are also less harmful to the environment, beneficial insects, natural enemies, and animals. Read all labeling, including the Safety Data Sheet, before making your final decision to ensure that you are working with the least dangerous and most effective pesticide for your situation.

Pesticide Formulations

Sometimes you must choose between two or more formulations of the same pesticide to control a target pest. When possible, make your selection based on the type of control desired, safety, cost, and other factors, such as those listed in Table 11-1. For example, emulsifiable formulations of insecticides usually provide quick control but are active for less time than wettable powders. Whenever you have a choice, consider the safety of pesticide applicators and fieldworkers, as well as known environmental issues. For example, you should select a formulation with a low volatile organic compound (VOC) content whenever possible to reduce air pollution.

Evaluate the habits and growth patterns of each pest. Be sure the formulation is right for the pest's life stage. Pick a formulation that affects the environment the least. Consider drift, runoff,

TABLE 11-1:

Comparison of pesticide formulations

Formulation	Mixing/loading hazards	Phytotoxicity	Effect on application equipment	Agitation required	Visible residues	Compatibility with other formulations
wettable powders (W, WP)	dust inhalation	low	abrasive	yes	yes	excellent
dry flowables/ water-dispersible granules (DF or WDG)	low	low	abrasive	yes	yes	good
soluble powders (SP)	dust inhalation	usually low	nonabrasive	no	some	fair
emulsifiable concentrates (EC or E)	spills and splashes	maybe	may affect rubber pump parts	yes	no	fair
flowables (F) and soluble concentrates (SC)	spills and splashes	maybe	may affect rubber pump parts; abrasive	yes	yes	fair
solutions (S)	spills and splashes	low	nonabrasive	no	no	fair
dusts (D)	severe inhalation hazards	low	—	yes	yes	—
granules (G) and pellets (P)	low	low	—	no	no	—
microencapsulated formulations	spills and splashes	low	—	yes	—	fair

wind, and rainfall, along with soil type and characteristics of the surrounding area. Finally, choose a formulation that will work best with your application equipment.

Pesticide Persistence

Depending on your pest problem, consider persistence when choosing a pesticide. Persistence is desirable in places where pest reoccur over time, such as for control of soil-dwelling pests. Persistence is important to consider when choosing herbicides, because residues may damage the next crop.

The pH of the water used for mixing pesticides affects the breakdown speed. The pH of the soil or plant or animal tissues may have a similar effect. Tissue or soil that is highly alkaline (higher pH) often causes more rapid breakdown of some pesticides than neutral or acidic (lower pH) tissue or soil.

The physical nature of the treated surface also affects pesticide persistence. Porous surfaces or soils high in organic matter absorb pesticide, reducing the amount of a.i. available for pest control. Soil microorganisms also break down many pesticides, influencing the persistence of pesticides in the soil environment. Oily surfaces and waxy coatings on leaves and insect body coverings prevent uptake of the pesticide and may even combine with the a.i., reducing toxicity and persistence.

Water-soluble pesticides that soak deeply into the soil break down more slowly than those that remain near the surface, because there are fewer microorganisms in deep soil. High levels of organic matter in the soil often slow breakdown because the organic matter binds to the pesticide, making it unavailable to microorganisms. Repeated use of the same pesticide in the soil can increase the breakdown rate because the pesticide is good for the microorganisms or the microorganisms become more efficient breakdown agents.

Weather affects persistence. For instance, wind and rain blow or wash pesticides off target surfaces, making them less effective. High temperatures and humidity cause chemical changes that speed up the breakdown of some pesticides. Sunlight produces photochemical reactions that break down many pesticides. Cooler soil temperatures usually slow pesticide breakdown.

Cost and Efficacy of Pesticide Materials

The cost of a pesticide is an important factor, but be careful not to base your choice only on cost. A pesticide that costs 30% more but gives 60% better control is often the better bargain, unless it interferes with harvesting or other operations or you need to protect natural enemies.

Ease of Use and Compatibility with Other Materials

Pesticides that are easy to use and work well with other pesticides have an advantage. Compatibility and ease of use depend on

- how the pesticide is being used
- what the pesticide is mixed with
- the nature of the treatment area

Effects on Beneficial Insects and Natural Enemies

Always try to protect beneficial insects and natural enemies. If you are using an integrated pest management program, consider how the pesticide will work within the goals of the program. For example, it may be better to settle for slower control of the pest if the result is greater long-term control.

Restricted-Entry and Preharvest Intervals

The pesticide you choose must be able to be applied in a way that meets both the legally established restricted-entry intervals and the allowable days before harvest.

Applying Pesticides Effectively

Pesticide Application Methods

Use specific pesticide application methods to improve coverage, reduce drift, and achieve better control of pests. Choosing the right application methods can also reduce human and environmental hazards.

Equipment Operation. Learn how to run pesticide application equipment properly. For example, you must keep ground speed the same to ensure even application across a field unless you are using a rate-controlling system. Check nozzles frequently to make sure they have not gotten clogged and that the spray pattern remains even. Shut off all nozzles during turns to keep spray patterns even and to avoid contaminating nontreatment areas. When injecting pesticides into the soil, shut off nozzles and raise the boom before making a turn. Leave enough room after the turn to bring equipment up to the specified ground speed before restarting the pesticide flow.

Explain how a GPS unit can impact the effectiveness of pesticide applications.

Preventing Gaps or Overlaps. Pesticide swaths must be uniform, without overlaps or gaps, to make the best economic use of spray materials. In some agricultural settings, you can follow furrows or rows to keep the application uniform. In open, unmarked areas, you must depend on some other method to prevent overlap or gaps. One method involves using foam markers to mark sprayed areas to prevent overlap or gaps in the application pattern. In some situations, you can add a colored dye to the spray mixture to show where spraying has taken place. Markers can also be used to mark the exact point in a field where you stopped spraying (for times when you leave the field to refill the spray tank, for example). You can also use electronic positioning devices, such as global positioning systems (GPS), to guide pesticide application equipment back to the exact point where spraying stopped (Fig. 11-3), which helps prevent gaps and overlaps.

FIGURE 11-3.

A GPS unit mounted on your sprayer will guide you back to the exact spot where spraying stopped.

Application Timing

Proper application timing is important for controlling target pests as well as protecting natural enemies and beneficial insects. Because some pesticides are more effective at different life stages of the pest, time applications to control the most susceptible stage. Understanding the biology of the pest will help you determine its susceptible life stages and decide whether a pesticide application will work. You must know a pest's habits, that is, when it is most active and likely to be affected by a pesticide application. Equally, you must know the habits of honey bees and other nontarget organisms so you can avoid the consequences of damaging the ecosystem, such as pest resurgence. See "Protecting Nontarget Organisms" later in this chapter for more about timing applications to avoid bees and other beneficial insects.

Mode of Action and Pesticide Uptake

Knowing a pesticide's mode of action will help you choose the best way to apply it so that it reaches the intended pest and protects other organisms. To ensure that the pesticide is taken up by the pest, you must also understand where the pest lives and feeds so you can deliver the pesticide to the correct place. Considering a pesticide's mode of action is also important when you are faced with resistant pests. You can delay pesticide resistance in target organisms by rotating through pesticides with differing modes of action. For more thorough information about mode of action, see Chapter 3.

Mixing Pesticides

Explain how to determine whether two or more pesticides will be compatible for tank mixing.

Describe mixing procedures for

- a single pesticide
- two or more pesticides (tank mix)

Mixing pesticides requires focus and attention to detail. Forgetting a step in the process or using too much (or too little) of the formulation can cause costly and sometimes dangerous problems. Methods to help you mix pesticides safely and effectively are explained below. You will also find resources you can use to find out whether the chemicals you are mixing are compatible.

Effective Methods

Before preparing a tank mix, be sure the spray tank is thoroughly clean and contains no sediments or residues.

Because mixtures of pesticides may be incompatible and may separate or curdle, you must test for compatibility before preparing a full tank. You can evaluate the tank mixture by performing the simple compatibility test described in Sidebar 11-1.

Some labels will provide pesticide mixing order and procedures. However, if you cannot find specific instructions for the materials you are mixing, you can follow these general guidelines:

1. Add some of the diluent (usually water) and then add some adjuvants (like surfactants) one at a time.
2. Add wettable and other powders and water-dispersible granules one at a time.
3. Agitate thoroughly and add the remaining diluent.
4. Add liquid products (like flowables and certain adjuvants) and water-soluble concentrates one at a time.
5. Add emulsifiable concentrates one at a time.

For example, when combining a water-soluble concentrate with a wettable powder, always add the wettable powder first. When mixing an emulsifiable concentrate with a dry flowable, add the dry flowable first.

Field Incompatibility

Sometimes tank mixes seem compatible during testing and after mixing in the spray tank but stop working well together during application. This problem is called field incompatibility. The temperature of the water in the tank can cause this problem. It could also be due to water impurities. Sometimes the amount of time the spray mixture has been in the tank causes field incompatibility. From time to time, variations among different lots of pesticide chemicals are great enough to cause this type of incompatibility.

Resolving Compatibility Problems in the Spray Tank

Try the following things if pesticide incompatibility develops in the spray tank. First, increase agitation and try to break up clumps with a water stream to get the mixture recirculating. If the material still separates, contact your pesticide dealer for an effective compatibility agent. Add the agent to the tank and continue agitation.

Changing filter screens to a larger size and cleaning them often may help eliminate some clumping. If these steps do not resolve the problem, add additional water to the mixture and filter off larger particles. If you cannot spray the mixture onto an application site, put it into an appropriate container for disposal. Follow the same procedures you would use to dispose of any other unused pesticide.

SIDEBAR 11-1

Compatibility Test for Pesticide Mixtures

WARNING

Always wear the label-required PPE when pouring or mixing pesticides. Perform this test in a safe area away from food and sources of ignition. Pesticides used in this test should be put into the spray tank once testing is completed. Rinse all utensils and jars, and pour rinsate into the spray tank. Do not use utensils or jars for any other purpose after they have contacted pesticides.

TEST PROCEDURE

1. Measure 1 pint of the intended spray water into a clear quart glass jar.
2. Adjust pH if necessary (see Sidebar 11-2).
3. Add ingredients in the following order. Stir well each time an ingredient has been added.

Material to be added	Amount of material to add per 100 gallons of spray mixture
1. Surfactants, compatibility agents, and activators	1 teaspoon for each pint
2. Wettable powders and dry-flowable formulations	1 tablespoon for each pint
3. Water-soluble concentrates or solutions	1 teaspoon for each pint
4. Soluble powder formulations	1 teaspoon for each pint
5. Remaining adjuvants	1 teaspoon for each pint
6. Emulsifiable concentrate and flowable formulations	1 teaspoon for each pint

7. After mixing, let the solution stand for 15 minutes. Stir well and observe the results.

TEST RESULTS

compatible	Smooth mixture, combines well after stirring. Chemicals can be used together in the spray tank.
incompatible	Separation, clumps, grainy appearance. Settles out quickly after stirring. Follow instructions below to try to resolve incompatibility; otherwise do not mix this combination in the spray tank.

RESOLVING INCOMPATIBILITY

1. Add 6 drops of compatibility agent and stir well. If mixture appears compatible, allow it to stand for 1 hour, stir well, and check it again. If the mixture appears incompatible, repeat one or two more times, using 6 drops of compatibility agent each time.
2. If incompatibility still persists, dispose of this mixture, clean the jar, and repeat the above steps, but add 6 drops of compatibility agent to the water before anything else is added.
3. If the mixture is still incompatible, do not mix the chemicals in the spray tank. To overcome this problem, you might consider the following alternatives:
   - Use a different water supply.
   - Change brands or formulations of chemicals.
   - Change the order of mixing.
4. Make only one change at a time, and perform a complete test, as described above, before making another change. Do not mix the chemicals in the spray tank if incompatibility cannot be resolved.

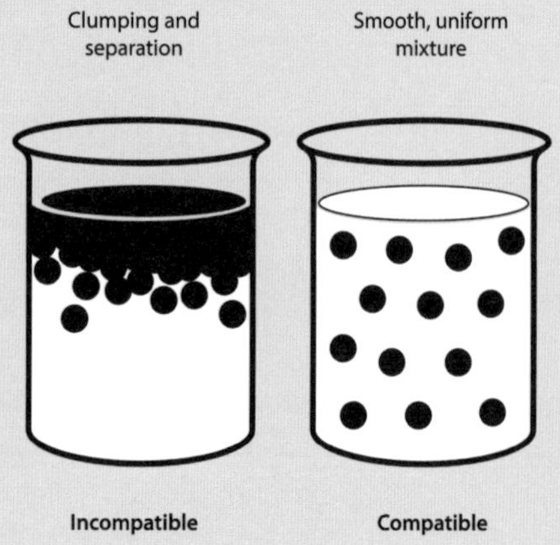

SIDEBAR 11-2

Testing and Adjusting pH of Water Used for Mixing Pesticides

You can measure pH with an electronic pH meter, a pH test kit such as those used for testing swimming pool water, or pH test paper available from a chemical supply dealer.

TEST WATER

1. Using a clean container, obtain a sample of water from the same source that will be used to fill the spray tank.
2. Measure exactly 1 pint of this water into a clean quart jar.
3. Check the pH of the water using a pH meter, test kit, or test paper.

pH LEVEL

3.5–6.0: Satisfactory for spraying and short-term (12- to 24-hour) storage of most spray mixtures in the spray tank.

6.1–7.0: Adequate for immediate spraying of most pesticides. Do not leave the spray mixture in the tank for more than 1 to 2 hours, to prevent loss of effectiveness.

Above 7.0: Add a buffer or acidifier.

ADJUST pH

1. Using a standard eyedropper, add 3 drops of buffer or acidifier to the measured pint of water.
2. Stir well with a clean glass rod or other clean, nonporous utensil.
3. Check pH as above.
4. If further adjustment is needed, add 3 drops of buffer or acidifier, stir well, then recheck pH. Repeat until pH is satisfactory. Remember how many times 3 drops were added to bring the solution to the proper pH.

CORRECT pH IN SPRAY TANK

1. Before adding pesticides to the sprayer, fill the tank with water.
2. For every 100 gallons of water in the spray tank, add 2 ounces of buffer or acidifier for each time 3 drops were used in the jar test above. Add buffer or acidifier to water while agitators are running. If tank is not equipped with an agitator, stir or mix well.
3. Check pH of the water in the spray tank to be certain it is correct. Adjust if necessary.
4. Add pesticides to spray tank.

Chemical Changes with Pesticides and Pesticide Combinations

In some tank mixes, pesticides may mix properly in solution but the effectiveness or toxicity of the pesticides in the mixture changes. These changes are due to chemical, rather than physical, reactions between combined pesticides, impurities, tank linings, or the water used for mixing. Such changes are hard to recognize because you cannot see them, so read labels carefully to find out about chemical reactions that can happen during mixing.

Pesticide Resistance

Pesticide resistance is a genetic trait a pest individual inherits that allows it to survive an application of a pesticide at label rates that kills most other individuals in the population. After surviving the pesticide application, the resistant individual then passes the gene(s) for resistance on to the next generation. The more the pesticide is used, the more susceptible individuals are eliminated and the larger the number of resistant individuals grows until the pest population is no longer effectively controlled (Fig. 11-4).

Pesticide resistance is most common in arthropods, with over 500 species of resistant insects and mites reported worldwide. However, pesticide resistance is increasing among other types of pests as well. Certain populations of bacteria, fungi, vertebrates, and weeds are becoming resistant to greater numbers of pesticide products. To find out more about pests that are known to have become resistant to pesticides, check the following websites (according to pest type):

- Insecticide Resistance Action Committee, irac-online.org/
- Fungicide Resistance Action Committee, frac.info/
- Herbicide Resistance Action Committee, hracglobal.com/
- Weed Science Society of America, wssa.net/wssa/weed/herbicides/

The most important method you can use to fight the development of resistance is monitoring. Monitoring helps you

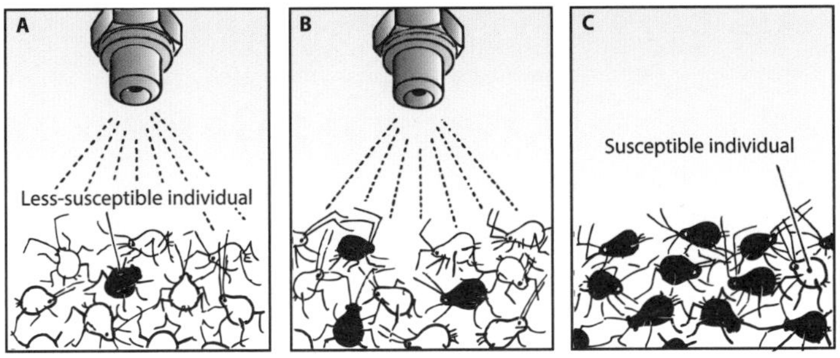

FIGURE 11-4.

Pest populations develop resistance to pesticides through genetic selection. (A) Certain individuals in a pest population are less susceptible to a pesticide spray than other individuals. (B) These less-susceptible pests are more likely to survive an application and to produce less-susceptible progeny. (C) After repeated applications, the pest population consists primarily of resistant or less-susceptible individuals, and applying the same material or other chemicals with the same mode of action is no longer effective.

detect tolerance to pesticides in pest populations before resistance becomes widespread. Early detection provides the opportunity to integrate other pest management options to prevent or reduce selection for resistance. For example, combining different weed control practices such as increased cultivation and crop rotation with the use of herbicides can reduce the number of resistant weed biotypes. By alternating pesticide treatments with different modes of action, it may be possible to use some pesticides for a longer period of time.

Factors Influencing Selection for Resistance

List the factors that contribute to pesticide resistance.

Generally, the development of pesticide resistance is similar for insects, mites, fungi, bacteria, weeds, and vertebrates. It involves a combination of genetic, biological, and other operational factors.

Genetic Factors. Genetic factors that influence the development of resistance include

- how the organism inherits resistance
- how many of the individuals in the population have the genes for resistance

Individuals with the resistance gene are able to tolerate that pesticide and will be able to survive and reproduce. Individuals without the resistance gene cannot tolerate the pesticide and will die. If resistance genes are common, the population inherits them easily. Resistance will spread quickly and may be hard to manage.

Once a pest is resistant to one pesticide, resistance to others may follow more quickly. This phenomenon, called cross-resistance, happens when the pest is resistant to two or more pesticides at the same time. Multiple resistance happens when pests have several distinct mechanisms to withstand pesticide chemicals, allowing them to tolerate several groups of pesticides that are not chemically related to each other.

Biological Factors. The biology of the pest species influences the rate at which resistance will happen. Biological characteristics include the pest's life span, reproductive capabilities, and mobility. Typically, a short-lived, fast-developing, immobile pest population that produces many offspring will develop resistance quickly. Resistance happens more slowly when there are untreated refuges available or when the pest species (insect, pathogen, or vertebrate) is highly mobile. In weeds, resistance happens more quickly when the plant produces lots of seeds that are likely to sprout.

Operational Factors. Operational factors can be controlled by people. Those that favor resistance include the pesticide's type, persistence, mode of action, and application method (the rate applied, frequency, whether it was mixed with other pesticides, and timing in relation to the dynamics of the pest population).

Management decisions involving these factors can increase or reduce resistance. Repeated use of a single pesticide, for example, increases the risk of resistance, especially when other control methods are not used.

Preventing Offsite Movement of Pesticides

Describe the different types of offsite movement, including factors that can affect the occurrence of each type.

Preventing Pesticide Drift

Drift can be defined as the airborne movement of pesticides to nontarget areas. This type of offsite movement can be in the form of spray droplet drift, vapor drift, or particle (dust) drift. Studies have shown that a high percentage of pesticides may never reach the target site because of drift, so you must be on the lookout for conditions that favor drift. By avoiding these conditions, it is possible to reduce drift to a tolerable level (Table 11-2). The most important factors affecting drift include

- wind speed and direction
- droplet size
- applicator (liquid or dust) proximity to the edges of the treated area

- release height
- formulation type
- pesticide chemical characteristics

ASSESSING WEATHER CONDITIONS

The most accurate weather observations for a specific location will come from a weather station installed at that site. A computer-assisted weather station generally has an electronic data logger, sensing devices, power supply, environmental enclosure, and a support structure, but these can be expensive and hard to maintain. Stand-alone data loggers track weather conditions to help you predict pest problems. These handheld, battery-operated loggers do not have to be attached to a computer to measure and accumulate minimum-maximum temperatures. Be sure to set up and maintain the weather instruments according to manufacturers' instructions and calibrate them

TABLE 11-2:

How to avoid various types of drift

Type of drift	Variables	Reduction measures
spray drift	Air movement: • Wind above 10 mph carries small droplets away from the application site.	Create larger droplets by • using nozzles with a larger orifice • using Venturi or air-induction nozzles • using a lower pressure setting when spraying • using a thicker pesticide formulation (such as an invert emulsion) • using deposition aids or thickeners Avoid applying pesticides when wind speed is greater than 10 mph.
	Temperature: • High temperatures can increase the likelihood of drift via evaporation. • Varying temperatures may result in an inversion condition that can carry concentrated clouds of pesticide long distances.	Avoid applications • during midday, when temperatures near the ground increase • during temperature inversions, when the air temperature over 20 feet above the ground is warmer than the air below
	Low relative humidity: • Especially when combined with high temperatures, this condition increases evaporation of water used to carry pesticide, creating smaller droplets.	Avoid applications when humidity is low, especially when temperatures are also high. Apply instead during morning or evening hours, when temperatures are lower and humidity is likely to be higher.
	Distance from target: • The longer a droplet has to travel to reach the target site, the more likely it is to drift.	Adjust nozzle/boom height to shorten the distance a pesticide must travel to reach the target.
vapor drift	Temperature: • High temperatures (above 85°F) increase the likelihood of volatilization.	Avoid applications during periods of high temperatures or when high temperatures are expected up to several hours after application. Inject volatile pesticides deep into moist, packed, or tarped soil, or treat the site with water immediately following the application. Choose a low-volatility formulation and/or a pesticide with a low vapor pressure.
	Equipment type: • Broadcast applications can increase the potential for pesticides to volatilize.	Applications using a chemigation drip system can reduce volatilization potential.
particle (dust) drift	Air movement: • Wind can blow pesticide-contaminated soil particles away from an application site.	Avoid applying persistent pesticides to soil in areas subject to high winds. Keep soil moist if high winds are expected after an application of persistent pesticides that are likely to bind to soil particles.
	Equipment type: • Air blast sprayers used in orchards can blow pesticide-contaminated soil into the air.	Avoid using air blast sprayers after orchard floors have been treated with certain herbicides.

FIGURE 11-5.

Surveyor's tape hung from a tree branch is an indicator of relative wind speed and direction, helping you make the best possible application decision.

regularly to ensure accuracy. Dirty, poorly sited, or poorly maintained instruments will give inaccurate measurements that can lead to bad decisions.

Alternatives to expensive computer-aided weather assessment tools and electronic handheld devices include

- estimating wind speed by observing a strip of surveyor's tape hung from a tree branch or pole (Fig. 11-5), watching wind movement through trees, or using a wind sock
- gathering accurate temperature readings by listening to local radio stations or NOAA Weather Radio and VHF channels, or using various weather resources online, including the UC IPM California Weather Database (see the UC IPM website, ipm.ucanr.edu)

Spray Drift

Spray drift occurs more often than vapor or dust drift because spray applications are common and will almost always create some offsite movement. You must learn how to reduce the likelihood that sprays move away from the application site.

Avoid most problems associated with spray drift by paying close attention to spray droplet size, wind direction and speed, and boom or nozzle height. Larger spray droplets are less likely to drift than smaller ones, and droplets released close to the target site are more likely to stay on target. Typically, larger nozzle orifices and lower pressures produce larger droplets (Table 11-3). However, some new nozzles, such as the Venturi, or air-induction, nozzles produce larger droplets.

TABLE 11-3:

American Society of Agricultural and Biological Engineers (ASABE) Standard characteristics of spray droplets with nozzle catalog code, categories, symbols, and approximate micro sizes and relative comparisons

ASABE Standard				Comparative size*		
Symbol	Category	Code	Approximate VMD†	Relative size	Comparative size	Atomization
VF	very fine	red	< 100		point of needle (25 microns)	fog
F	fine	orange	100–175		human hair (100 microns)	fine mist
M	medium	yellow	175–250		sewing thread (150 microns)	fine drizzle
C	coarse	blue	250–375		fishing line (250 microns)	light rain
VC	very coarse	green	375–450		staple (420 microns)	rain
XC	extremely coarse	white	> 450		#2 pencil lead (2,000 microns)	thunderstorm

Notes: *Fine droplets are more likely to drift than coarse droplets; coarse droplets are more likely to run off. Selecting the correct nozzle will make your application more effective.

†Volume median diameter.

The viscosity (thickness) of the spray affects droplet size. Viscosity is a measure of a liquid's resistance to flow: for example, mayonnaise is more viscous than water. As the viscosity of the liquid increases, so does the droplet size, reducing the potential for offsite movement. Formulations such as invert emulsions have a thick consistency that aids in reducing drift.

Some drift control additives (adjuvants) can help reduce the potential for drift even after you have switched to drift-control nozzles. Remember, always follow the label directions when using any spray adjuvant intended for minimizing drift.

Air movement is the most important environmental factor influencing the drift of pesticides from target areas. The movement of air is influenced by the temperature at ground level and the temperature of the air above it, so taking weather readings can help you decide when drift is less likely to occur. Except in the case of temperature inversions (Fig. 11-6), the early morning and evening are often the best times to apply pesticides. This is because windy conditions are more likely to occur around midday when the temperature near the ground increases.

Droplets that travel shorter distances through the air to the target are less likely to drift. Carefully adjusting nozzle height helps reduce the chances that the pesticide will move away from the application site through the air.

Low relative humidities, high temperatures, or a combination of both, especially when spraying fine droplets, can also increase the potential for spray drift. Under these conditions, the evaporation rate of water increases, resulting in even smaller spray droplets that drift more easily. Avoid spraying during these times, or, if you must spray, switch to using nozzles that produce larger droplets.

FIGURE 11-6.

A temperature inversion is caused by a layer of warm air occurring above cooler air close to the ground. This warm air prevents air near the ground from rising, similar to a lid.

Reduce outdoor drift problems by spraying when the wind speed is low, by leaving an untreated border or buffer area in the downwind target area, and by spraying downwind from sensitive areas such as residential properties, schools, crops, waterways, or beehives. Be sure to adjust the height of nozzles so that they are spraying pesticide as close to the target as possible while still maintaining proper coverage. Using low-volatile or nonvolatile pesticides and using only low-pressure treatments can reduce indoor pesticide drift problems in places such as covered livestock pens, stables, or henhouses.

Restricted Use Pesticide

(Ground and Surface Water Concerns)

For retail sale to and use only by certified applicators or persons under their direct supervision and only those uses covered by the certified applicator's certification. This product is a restricted use herbicide due to ground and surface water concerns. Users must read and follow all precautionary statements and instructions for use to minimize potential for Atrazine to reach ground and surface water.

FIGURE 11-7.

A pesticide label statement warning users that this material is restricted because of groundwater and surface water concerns.

Preventing Surface Water and Groundwater Contamination

To help prevent surface water and groundwater contamination, the U.S. EPA requires that all pesticide products with directions for outdoor uses must include the environmental hazard statement on the label: "Do not apply directly to water, or to areas where surface water is present, or to intertidal areas below the mean high water mark. Do not contaminate water supplies when cleaning equipment or disposing of equipment wash waters." Pesticides that have the potential to be found in groundwater must bear groundwater warning statements on their labels (Fig. 11-7). Groundwater statements on labels help you choose appropriate pesticides where soils are sandy or where extra precautions are needed to reduce contamination risk. Also, check to make sure you are not in a Ground Water Protection Area before applying any pesticide that appears on California's Groundwater Protection List.

You can reduce the risk of contaminating surface water and groundwater by following the practices listed below.

Use IPM principles. Apply pesticides only when and where necessary and only in amounts adequate to control pests. Following IPM principles, use nonchemical control methods whenever possible.

Identify vulnerable areas. The presence of sandy soil, sinkholes, wells, streams, ponds, and shallow groundwater increases the chance of groundwater contamination. Avoid pesticide application in these locations, if possible. Never dispose of empty pesticide containers in sinkholes or dump or rinse sprayers into or near sinkholes. Do not under any circumstances clean tanks or intentionally discharge water from the tank of any vehicle into a street, along a road, or into a storm drain. For up-to-date lists and maps of Ground Water Protection Areas in California, see the DPR website, cdpr.ca.gov/docs/emon/grndwtr/gwpa_locations.htm.

Do not mix and load near water. Carry out mixing and loading as far as possible (at least 100 feet) from wells, lakes, streams, rivers, and storm drains. When possible, mix and load pesticides at the site of application. Consider using a sealed permanent or portable mixing and loading pad to prevent contaminating soil.

Keep pesticides away from wells. Do not store or mix pesticides around wells. Poorly constructed or improperly capped or abandoned wells can allow surface water containing pesticides and other contaminants direct entry into groundwater.

Avoid back-siphoning. Back-siphoning is the reverse flow of liquids into a fill hose. It sucks tank contents back into the water supply. Back-siphoning starts with a reduction in water pressure and can draw very large quantities of pesticide directly into the water source. This happens when the end of the water hose is allowed to extend below the surface of the spray mixture when filling a spray tank. The simplest method of preventing backflow is to maintain an air gap between the discharge end of the water supply line and the pesticide solution in the spray tank. Keep the air gap at least twice the diameter of the discharge pipe. Another method for preventing back-siphoning is to use an anti-backflow device or check valve.

Improve land use and application methods. Terraces and conservation tillage can reduce water runoff and soil erosion. Ideally, leave as much plant residue as possible on the soil surface to keep erosion levels low. Where conservation tillage is not possible, reduce runoff potential by incorporating pesticides into the soil. This practice lowers the concentration of pesticide on the soil surface.

Grass buffer strips are very effective in reducing pesticide runoff because they trap sediment containing pesticides and slow runoff water, allowing more runoff water to infiltrate the soil. Leaving untreated grass strips next to streams, ponds, and other sensitive areas can trap much of the pesticide running off treated areas.

Time pesticide applications according to the weather forecast. Pesticides are most likely to move away from an application site during heavy rains or irrigation during the first several hours after application. Choose products wisely. Whenever possible, use pesticides that are less likely to leach. Read labels for leaching warnings.

Protecting Sensitive Areas

Sometimes pesticides must be deliberately applied to a sensitive area to control a pest. Only applicators who are competent in handling pesticides should perform these applications. At other times, the sensitive area may be part of a larger target site. Whenever possible, take special precautions to avoid application to the sensitive area. Leaving an untreated buffer zone around a sensitive area is a good way to avoid contaminating it. In still other instances, the sensitive area

Sensitive Areas: The pesticide should only be applied when the potential for drift to adjacent sensitive areas (e.g., residential areas, bodies of water, known habitat for threatened or endangered species, non-target crops) is minimal (e.g., when wind is blowing away from the sensitive areas).

FIGURE 11-8.

A pesticide label statement describing the restrictions on using this material near sensitive areas.

may be near a site used for mixing and loading, storage, disposal, or equipment washing. In all cases, you must take precautions to avoid accidental contamination of sensitive areas. Check the label for statements that alert you to special restrictions around sensitive areas, such as the one in Figure 11-8. There are additional restrictions in California when applying near schools. Check with your local agricultural commissioner and in the 3 CCR list for details.

Protecting Nontarget Organisms

The following sections discuss some methods you can use to minimize pesticides' effects on nontarget plants; bees and other beneficial organisms; and fish, wildlife, and livestock.

Nontarget Plants

Pesticides can cause plant injury due to chemical exposure (phytotoxicity), particularly if they are applied at too high a rate, at the wrong time, or under unfavorable environmental conditions. Check the pesticide label carefully to make sure you have calculated the application rate correctly and are using the right application equipment. You must also plan applications ahead of time to make sure that conditions are ideal and that both target and nontarget organisms are at the optimal life stage. For instance, you may want to delay applying herbicides to a crop if it has recently been water stressed, since weakened plants are more likely to experience phytotoxic effects. Use the calibration methods in Chapter 10 to ensure that application rates and coverage areas remain constant. Most unintended phytotoxic injury is due to herbicides that drift into adjacent areas, though it may sometimes be caused by surface runoff. See earlier paragraphs in this chapter covering drift and runofffor methods you can use to avoid these types of offsite movement.

Bees and Other Beneficial Organisms

Because bees are such an important part of agricultural ecosystems, you must be aware of bee activity when planning pesticide applications. Preventing bee loss is the joint responsibility of the applicator, the grower, and the beekeeper. Check with your local county agricultural commissioner before performing pesticide applications that may harm bees. The commissioner will have contact information for beekeepers who have requested notification prior to pesticide applications, as well as the specific bee protection regulations or unique county conditions that can impact your application. You can minimize losses of bees to insecticide poisoning by following a few basic principles.

- Read the label and follow label directions.
- Determine whether bees are foraging in the target area so you can take protective measures.
- Whenever possible, use pesticides and formulations least hazardous to bees. Emulsifiable concentrates are safer than powders and dust formulations. Granules are the safest and least likely to harm bees. Microencapsulated pesticides pose the greatest risk to bees, as do all types of systemic insecticides.
- Choose the least hazardous application method. Ground applications of certain insecticides are less hazardous to bees than aerial applications. All methods of application are hazardous to bees when applying systemic insecticides.
- Apply chemicals in the evening or during early-morning hours before bees forage. Evening applications are generally safer to bees than morning applications. If unusually warm evening temperatures cause bees to forage later than usual, delay the pesticide application.
- Do not spray crops in bloom except when necessary; never spray when bees are present.
- Do not spray when weeds or other plants around the treatment site are in bloom.
- Do not treat an entire field or area if spot treatments will control the pest.

Table 11-4 lists pesticides according to their impact on honey bees. If a pesticide can harm honey bees, it can also cause problems for other beneficial organisms. Often these beneficial organisms are valuable allies in keeping pest populations below damaging levels. A pesticide application can harm the beneficial organism population as much as the target pest, so do not spray when beneficial organisms are in the target area.

Fish, Wildlife, and Livestock

Fish kills often result from water pollution by a pesticide and are most likely to be caused by insecticides, especially when small ponds or streams are under conditions of low water flow or volume. Avoid situations where the pesticide you are applying can easily move into water or away from the application site in flowing water. For example, applications to unterraced areas can expose animals below to pesticides moving in water that flows downhill. Read the label to make sure that you leave the required buffer zone between the application site and lakes, streams, or underground water supplies that may feed into bodies of water that support aquatic organisms. To find out more about how to minimize runoff and leaching, see "Preventing Surface Water and Groundwater Contamination" earlier in this chapter.

Bird kills resulting from pesticide exposure can happen in a number of ways. Birds may ingest pesticide granules, baits, or treated seeds; they may be exposed directly to sprays; they may consume treated crops or drink contaminated water; or they may feed on pesticide-contaminated insects and other prey. Granular or pelleted formulations are a particular concern because birds

TABLE 11-4:

Active ingredients of commonly used pesticides and their effect on bees

Active ingredient*	Pesticide type	Highly toxic to bees or brood	Toxic to bees or brood	No bee precautionary statement on label
abamectin	miticide	●		
azadirachtin	insecticide		●	
Bacillus thuringiensis ssp. *aizawai*	insecticide		●	
bifenthrin	miticide, insecticide	●		
chlorpyrifos	insecticide	●		
copper sulfate	bactericide, fungicide			●
diatomaceous earth	insecticide		●	
diazinon	miticide, insecticide	●		
fipronil	insecticide	●		
glyphosate	herbicide		●	
horticultural oil	insecticide, miticide, fungicide		●	
imidacloprid	insecticide	●		
mefenoxam	fungicide			●
oxyfluorfen	herbicide			●
paraquat	herbicide		●	
propiconazole	fungicide		●	
pyrethrin	insecticide	●		
rotenone	insecticide, piscicide		●	
spinosad	insecticide		●	
sulfur	miticide, fungicide			●

Note: *Some a.i.'s or products listed here may not be currently registered as pesticides or may have had their registration cancelled.
Source: Adapted from Hooven et al. 2013.

and other animals often mistake them for food. Other formulations (liquid) may be safer when birds and other wildlife are in or near the treated area. Place baits properly so they are inaccessible to pets, birds, and other wildlife, and use seed baits that are colored so that nontarget birds will avoid them.

Animals can also be harmed when they feed on plants or animals carrying pesticide residues. Pesticide residues remaining on or in the bodies of dead animals may harm predators. This is called secondary poisoning. Avoid this situation by promptly removing and properly disposing of dead animals from the treatment area. Check the pesticide label for statements about secondary poisoning.

Livestock can also be harmed by pesticide applications. The most common source of livestock poisoning by pesticides is through ingestion of contaminated feed, forage, and drinking water. Contamination often occurs as a result of improper or careless transportation, storage, handling, application, or disposal of pesticides, so be careful when working with pesticides around livestock.

Describe how to evaluate spray coverage and adjust application variables to change coverage as needed.

Follow-Up Monitoring

Follow up after every pesticide application to determine whether the application was successful. Sidebar 11-3 is a follow-up checklist. Begin by comparing the amount of pesticide actually used with the anticipated amount. This should vary by no more than 10%. If more or less pesticide was applied, determine the cause. Check sprayer calibration and tank-mixing procedures, and recalculate the size of the target area. Look for clogged or worn nozzles and wear or blockage in the sprayer pumping system.

Inspect the application site to make sure coverage was adequate and uniform. Wear PPE if necessary. Look for

- signs of pesticide runoff
- lack of penetration into dense foliage
- shingling or foliage that is clumping together
- uneven coverage from top to bottom of large plants

One way to determine whether coverage is uneven includes checking foliage for white residue after a wettable powder application has dried, or add a white sunscreen product that will leave a similar white residue. Another way to find out whether a sprayer has penetrated dense foliage, is to place water-sensitive paper cards in the treatment area before the application. If their color remains unchanged or they show only a few spots, you can adjust your application equipment to improve your results. Methods for adjusting a sprayer's application rates can be found in Chapter 10.

SIDEBAR 11-3

Pesticide Application Follow-Up Checklist

AMOUNT OF PESTICIDE USED

a. Calculated amount required for job: ____________________

b. Actual amount used: ____________________

c. Variation—divide (a) by (b), then multiply by 100. Subtract answer from 100 (answer should be between +10 and −10)

____

COVERAGE

a. Uniform ______ or Uneven ______

b. Runoff? Yes / No

c. Describe level of penetration into all areas: ____________

____

EFFECTIVENESS

a. Target pest controlled or reduced below economic injury level? Yes / No

b. Condition of natural enemies:

____

c. Secondary pest outbreak? Yes / No

PROBLEMS

a. Spotting or staining of surfaces? Yes / No

b. Injury to plants? Yes / No

c. Other: ____________________________________

____

COMMENTS

____

____

____

Chapter 11 Review Questions

1. Understanding the life cycle or stages of pests will help you to ____________.

☐ a. schedule pesticide applications without monitoring for pests
☐ b. choose the most effective pesticide to apply
☐ c. create an IPM program that requires no pesticides

2. What can you do to reduce an insecticide's effect on honey bees?

☐ a. Use the product when pests are very active, because honey bees are less active at that time.
☐ b. Apply early in the morning or in the evening, because honey bees are less active in the environment.
☐ c. Make an aerial application, because it is much safer whether honey bees are more active or less active.

3. Which pesticide property would make the material more likely to move with water in surface runoff?

☐ a. high solubility
☐ b. high adsorption
☐ c. high volatility

4. Put the following list of ingredients in the correct order for jar testing pesticides for compatibility.

☐ a. surfactants
☐ b. emulsifiable concentrates
☐ c. water-soluble concentrates
☐ d. wettable powder
☐ e. diluent

5. Match the type of offsite movement with a method you can use to combat it.

1. spray drift	a. Avoid applying on a hot day or when high temperatures are expected up to several hours after application.
2. vapor drift	b. Adjust boom height to shorten the distance a pesticide must travel to reach the target.
3. particle drift	c. Check Ground Water Protection Area maps, and avoid applications to these sites whenever possible.
4. runoff	d. Keep the ground moist if high winds are expected after an application of persistent pesticides to soil.
5. leaching	e. Leave grass buffer zones, especially when application sites are near streams, ponds, or other sources of surface water.

6. True or false?

☐ True ☐ False a. To measure a pesticide spray's penetration into thick foliage, like in trees, place pH-sensitive sponges in the treatment area and check their saturation levels after the application.

☐ True ☐ False b. Using a GPS unit on your sprayer will help reduce drift by ensuring that you are spraying droplets that are large enough to stay on target.

☐ True ☐ False c. The pesticide label is the best place to find out if one product can be successfully mixed with another in the same tank.

☐ True ☐ False d. One of the first things to look for when doing follow-up monitoring at an application site is indications that pesticide coverage was adequate and uniform.

☐ True ☐ False e. Pesticides are only harmful to adult bees and never affect the brood inside the hive.

7. Which of the following situations would increase the likelihood of pesticide resistance?

☐ a. You use two different pesticides to control a pest that reproduces only once per year.
☐ b. You repeatedly use the same pesticides at the same application site.
☐ c. You apply a pesticide that has little or no residual effect.

CHAPTER 12
PESTICIDE EMERGENCIES AND EMERGENCY RESPONSE

First Aid ... 174
If Pesticides Get on Your Skin or Clothing ... 176
If Pesticides Get in Your Eyes ... 177
If Pesticides Are Inhaled ... 177
If Pesticides Are Swallowed ... 178
Pesticide Leaks and Spills ... 179
How to Deal with a Pesticide Fire ... 182
How to Deal with Stolen Pesticides ... 182
Misapplication of Pesticides ... 183
Reviewing Emergency Response to Accidents ... 184
Chapter 12 Review Questions ... 185

Knowledge Expectations

1. Define first aid.
2. Explain the procedures to follow in getting emergency medical treatment for exposure episodes.
3. Describe how to set up and execute an emergency response plan.
4. Describe pesticide poisoning/overexposure symptoms and signs.
5. Describe how to identify heat-related illness and give first aid.
6. Describe where to find information about first aid for a person involved in a pesticide incident and explain what to do if
   a. you get pesticides on your clothing
   b. you get pesticides in your eyes
   c. you inhale pesticides
   d. you swallow pesticides
7. List the contents of a well-equipped decontamination facility, including components specific to different formulations.
8. List the contents of a pesticide spill kit.
9. Describe what to do when faced with a pesticide leak or spill.
10. Describe what to do when faced with a pesticide fire.
11. Describe what to do when a pesticide product has been stolen.
12. Describe how to respond to the misapplication of pesticides.
13. Explain why any incident should be reviewed.

First Aid

Define first aid.

First aid is the help you give a person exposed to pesticides before they receive emergency help from a medical professional. However, first aid is not a substitute for professional medical care. The "Precautionary Statements" section of each pesticide label provides specific first aid information.

Protect yourself when giving first aid to a person suffering from pesticide exposure. Avoid getting pesticides onto your skin. Do not inhale vapors. Do not enter a confined area to rescue a person overcome by toxic pesticide fumes unless you have the proper PPE, including respiratory equipment. Remember, the pesticide that affected the injured person can also injure you.

Get professional medical care at once for anyone who was exposed to a highly toxic pesticide or who shows signs of pesticide poisoning. Call an ambulance or transport the injured person to a medical facility for treatment, whichever is quicker. An injured person should never transport themselves to a medical facility. In addition to the first aid measures listed below, speed in obtaining medical care often controls the extent of injury. Provide medical personnel with complete information about the pesticide suspected of causing the injury, including the entire label and the associated SDS (Safety Data Sheet). Also explain how the person was exposed to the pesticide.

SIDEBAR 12-1

EMERGENCY NUMBERS FOR PESTICIDE ACCIDENTS AND SPILLS

WHEN PEOPLE HAVE BEEN EXPOSED TO PESTICIDES

- Dial 9-1-1 for emergency medical assistance. Notify the operator that the problem is a pesticide exposure. Provide an accurate location and information on the type of pesticide involved.
- After obtaining medical treatment for exposed persons, determine whether a spill has taken place. Follow the instructions below for a spill.
- Contact the nearest agricultural commissioner's office to report the incident at the website cdfa.ca.gov/exec/county/countymap/.

FOR PESTICIDE SPILLS ON STATE OR FEDERAL HIGHWAYS

- Notify the local office of the California Highway Patrol and your local fire department (dial 9-1-1). Inform the emergency operator that a pesticide spill has occurred; provide accurate location and type of pesticide.
- Contact CHEMTREC at 1-800-424-9300 for assistance in cleaning up a pesticide spill.
- Contact the California Office of Emergency Services. Usually a written report will need to be filed. See caloes.ca.gov/home or contact the main office in Sacramento:

  Governor's Office of Emergency Services
  3650 Schriever Ave
  Mather, CA 95655 (1-916-845-8510)
- Contact your local agricultural commissioner's office. Find the commissioner's telephone number for the county where the accident occurred at cdfa.ca.gov/exec/county/countymap/.

FOR PESTICIDE SPILLS ON LOCAL CITY OR RURAL ROADS OR ON PRIVATE LAND

- Contact the local police or sheriff and local fire department (dial 9-1-1). Inform the emergency operator that a pesticide spill has occurred. Provide accurate location and type of pesticide.
- Contact CHEMTREC at 1-800-424-9300 for assistance in cleaning up a pesticide spill.
- Report the spill to the California Office of Emergency Services (see caloes.ca.gov/home).
- Contact your local agricultural commissioner's office. Find the commissioner's telephone number for the county where the accident occurred at cdfa.ca.gov/exec/county/countymap/.

EMERGENCY RESPONSE PLANNING

A carefully thought-out emergency response plan is one of the most important tools you can have to prevent an emergency situation from becoming a catastrophic event. Consider the following guidelines when developing an emergency response plan and training employees on how to use it:

- Designate an emergency coordinator. This person must have the knowledge and authority to direct and manage employee responses to a pesticide emergency and to coordinate the efforts of local emergency response agencies such as firefighters, police, and paramedics.
- Maintain a list of emergency response agencies (Sidebar 12-1). Include names and telephone numbers of all response agencies you may have to call to assist in an emergency. Organize the list in the order to be called.

Explain the procedures to follow in getting emergency medical treatment for exposure episodes.

Describe how to set up and execute an emergency response plan.

- Post the name, location, and phone number of the nearest emergency medical care facility in a prominent place at the work site (or in the work vehicle).
- Include an outline with your calling list of the information to be passed along during an emergency notification call that contains the following:
  - name and callback number of the person reporting the incident
  - precise location of the incident
  - general description of what has occurred
  - the exact name, quantity, and classification of each chemical involved
  - the extent of any injuries
  - potential danger to the environment and persons living in the area
- Prepare a map of your facility to include with your emergency response plan. Show a layout of all chemical storage buildings and bulk storage tanks; access roads; main shut-offs for electricity, water, and gas; perimeter fencing that could hinder access to the pesticide storage facility; the location of fire alarms, firefighting equipment, and personal protective equipment (PPE); and drainage easements on the site. Provide emergency response agencies (in California, these are your county agricultural commissioner and your city or county fire department) with an updated copy of this map whenever changes are made at the facility (Fig. 12-1).
- Provide your emergency response agencies with an area map that shows your facility in relation to the surrounding area, so that firefighters, police, and paramedics do not waste time trying to find your facility.
- Keep a product inventory of the types and quantities of chemicals stored at your facility. Your emergency response plan should reflect peak-season storage. Important information in the product inventory includes product names, container volumes, and locations of containers in the storage facility. Also, you must always keep copies of pesticide labels and Safety Data Sheets.
- Maintain an updated list of suppliers who can provide additional equipment and materials that may be needed in the event of an emergency.

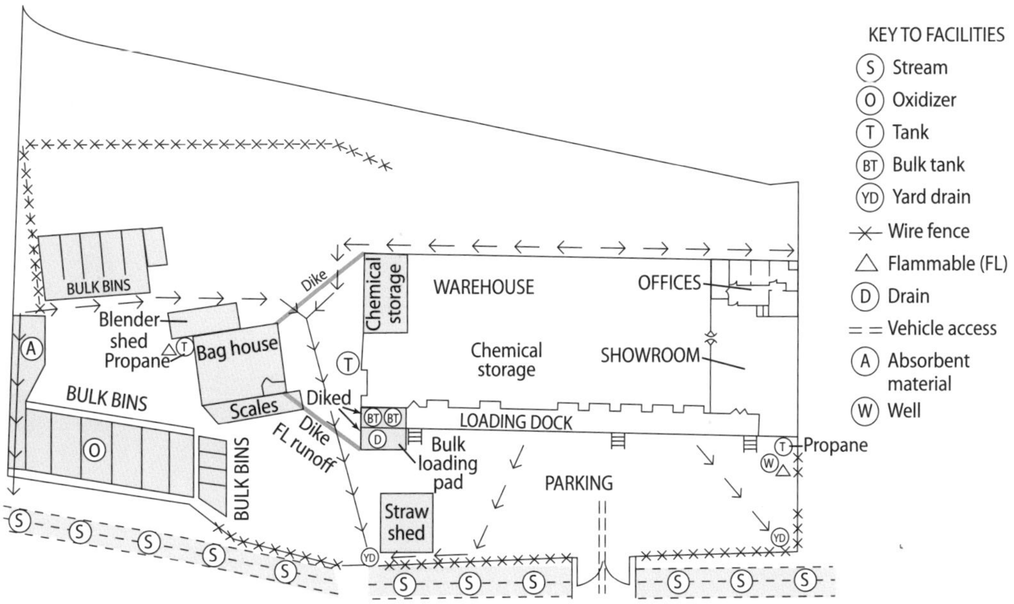

FIGURE 12-1.

Include a facility map as part of the emergency response plan. *Source:* Randall 2008.

Recognizing Pesticide Poisoning or Overexposure in People

Describe pesticide poisoning/overexposure symptoms and signs.

In order to provide the right first aid and communicate with medical personnel responding to the emergency, you must be able to recognize the signs and symptoms of acute pesticide poisoning or overexposure. Often symptoms of pesticide poisoning mimic symptoms of heat illness or common illnesses like the flu, so you must note what immediately preceded the onset of symptoms.

Poisoning or overexposure signs are what you can observe in a victim of pesticide poisoning. These can include vomiting, sweating, skin irritation, eye irritation, swelling, or pinpoint pupils.

Poisoning or overexposure symptoms can only be described by the victim, and may include nausea, headache, weakness, burning of the eyes, nose, mouth, or throat, chest pain, body aches, muscle cramps, and dizziness, among others.

Recognizing and Responding to Heat-Related Illness

Describe how to identify heat-related illness and give first aid.

Heat-related illness may mimic certain types of pesticide poisoning. Symptoms of heat illness include tiredness, weakness, headache, sweating, nausea, dizziness, and fainting. Severe heat illness can cause a person to act confused, get angry easily, or behave strangely.

Depending on the severity of the symptoms, first aid for heat-induced illness could encompass any or all of the following actions:

- Call 9-1-1 and notify a supervisor.
- Move workers to an air-conditioned or cool, shaded area where they can rest.
- Cool workers by
  - soaking their clothes with water
  - spraying, sponging, or showering them with water
  - fanning their bodies
- Provide workers with plenty of water, clear juice, or sports beverages to drink.

If Pesticides Get on Your Skin or Clothing

Describe where to find information about first aid for a person involved in a pesticide incident and explain what to do if
a. you get pesticides on your clothing
b. you get pesticides in your eyes
c. you inhale pesticides
d. you swallow pesticides

Pesticides that get on your skin or clothing can cause serious injury (Fig. 12-2). Some pesticides may cause skin burns or rashes or, through skin absorption, produce internal poisoning. Immediately remove contaminated clothing and wash the affected areas with clean water and soap.

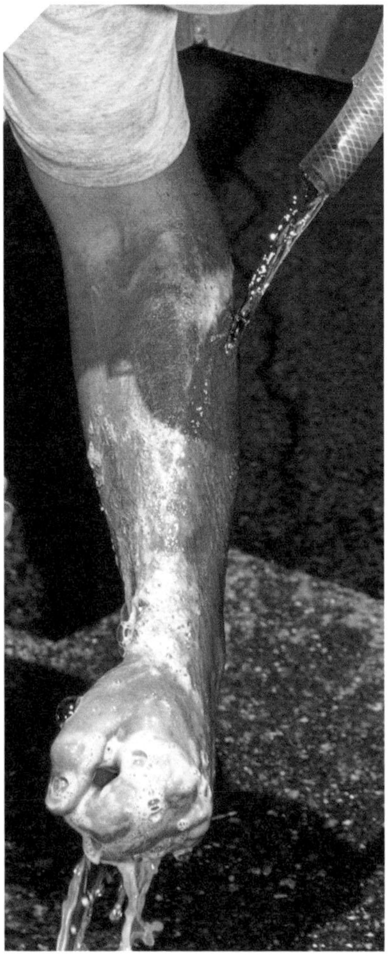

FIGURE 12-2.

If pesticides spill on you, the first step is to remove contaminated clothing and wash the affected parts of your body with soap and plenty of water. Do this quickly to avoid serious injury.

First Aid for Skin Exposure

Take the following actions if you or someone else receives skin exposure to pesticides.

- **Leave the contaminated area.** Get away (or remove the victim) from the fumes, spilled pesticide, and further contamination. Do this quickly!
- **Prevent further exposure.** Remove the contaminated clothing and thoroughly wash the affected skin and hair areas. Use soap or detergent and large amounts of water.
- **Get medical attention.** Call an ambulance or have someone transport the injured person to the nearest medical facility as quickly as possible. Give

medical providers a complete pesticide label and copy of the associated SDS of the pesticide that caused the injury.

If Pesticides Get in Your Eyes

Many pesticides can cause serious damage if they get into your eyes. Prompt first aid, followed by medical care, helps reduce damage.

FIGURE 12-3.

If you get a pesticide into your eyes, wash them with clean, running water for 15 minutes. Then, if irritation persists, seek medical care.

First Aid for Eye Exposure

Take the following actions to treat eye exposure:

- **Flush the eyes.** Immediately flush the affected eye or eyes with a gentle stream of clean, temperate, running water (Fig. 12-3). Direct the stream of water away from your eyeball, letting the water run off the bridge of your nose or from the side of your face and into your eye. Hold eyelids open to ensure thorough flushing. Continue flushing for at least 15 minutes (if you wear contacts, flush eyes for 5 minutes, then remove the contacts and continue flushing for 10 minutes). If only one eye is affected, keep the injured eye closer to the ground, so that the water used for flushing will not contaminate your other eye. When flushing your eyes, do not use any chemicals or drugs in the water, since this may increase the extent of injury. If running water is not available, slowly pour clean water from a glass, water cooler, or other container onto the bridge of your nose, rather than directly into your eyes.
- **Obtain medical care.** Always get medical attention if irritation persists after the flushing. Let medical providers know the name of the pesticide that caused the injury.

If Pesticides Are Inhaled

Inhaled chemicals, such as pesticide dusts, vapors from spilled pesticides, and fumes from burning pesticides, can cause serious lung injury and may be absorbed into other parts of the body through the lungs. Take first aid measures immediately to reduce injury or prevent death.

Wear an air-supplying respirator when entering an enclosed area to rescue a person who has been overcome by pesticide fumes. Cartridge respirators are not suitable for high concentrations of pesticide vapors or deficient oxygen conditions. If you do not have an air-supplying respirator, or have not been medically cleared to wear one, call for emergency help. You will be of more assistance to the injured person by seeking proper emergency help than if you are overcome by the pesticide fumes yourself.

First Aid for Pesticide Inhalation

Take the following actions if you need to provide first aid to someone overcome by pesticide fumes:

- **Leave the contaminated area, or remove an exposed person from the contaminated area.** Anyone overcome by pesticide vapors must get to fresh air immediately. Avoid physical exertion because this places an extra strain on the heart and lungs.

- **Loosen clothing.** Loosening clothing makes breathing easier and also releases pesticide vapors trapped between clothing and the skin.
- **Restore breathing.** If breathing has stopped or is irregular or labored, begin artificial respiration (rescue breathing). Continue assisting until breathing has improved or until medical help arrives. If the person has stopped breathing and has no pulse, begin cardiopulmonary resuscitation (CPR) and continue until help arrives.
- **Treat for shock.** Inhalation injury often causes a person to go into shock. Keep the injured person calm and lying down. Prevent chilling by wrapping the person in a blanket after removing contaminated clothing. Do not give them any alcoholic beverages.
- **Watch for convulsions.** If convulsions occur, protect the victim from falls or injury and keep air passages clear by making sure the head is tilted back.
- **Get immediate medical care.** Call an ambulance or take the person to the nearest medical facility. Give medical personnel as much information as possible about the pesticide.

If Pesticides Are Swallowed

Two immediate dangers are associated with swallowed pesticides. The first is related to the toxicity of the pesticide and the poisoning effect it will have on a person's nervous system or other internal organs. The second involves physical injury that the swallowed pesticide causes to the linings of the mouth and throat and to the lungs. Corrosive materials, those that are strongly acidic or alkaline, can seriously burn these sensitive tissues. Petroleum-based pesticides can cause lung and respiratory system damage, especially during vomiting. Never induce vomiting unless you have specific instructions to do so from a pesticide label or a medical professional. Always read the pesticide label to find out exactly what course of action is recommended.

You can reach regional poison information centers in Sacramento, San Francisco, Fresno, and San Diego by telephone 24 hours a day, 7 days a week. In a poisoning emergency, call the California Poison Control System (CPCS) by using the toll-free number 1-800-222-1222. These centers provide quick, lifesaving information on poisoning treatment. Having the label handy will help the operator react quickly and perhaps save lives.

First Aid for Swallowed Pesticides

Act quickly when a pesticide has been swallowed. Follow the pesticide label or poison information center instructions first. If you cannot access the pesticide label or CPCS, follow these guidelines:

- **Dilute the swallowed pesticide.** If the person is conscious and alert, give large amounts (1 quart for an adult or a large glass for a child under seven) of water or milk. Do not give any liquids to an unconscious or convulsing person.
- **Induce vomiting.** If the pesticide label indicates, induce vomiting. Make sure the person is kneeling or lying face down or on their right side. If you do not know what the label recommends, do not induce vomiting.
- **Obtain medical care.** Call an ambulance or take the poisoning victim to the nearest medical facility. Give medical personnel as much information as possible about the swallowed pesticide.

Setting Up Emergency Decontamination Facilities

List the contents of a well-equipped decontamination facility, including components specific to different formulations.

The decontamination process is a series of procedures performed in a specific order. For example, outer, more heavily contaminated items (e.g., outer boots and gloves) should be decontaminated and removed first, followed by decontamination and removal of inner, less contaminated items (e.g., jackets and pants). Each procedure should be performed at a separate station in order to prevent cross-contamination, if possible. The order of stations is called the decontamination line. Guidelines for the provision of safe, effective handler decontamination facilities can be found in Appendix D.

At a minimum, your decontamination facility must include the following items contained in chemical-resistant containers:

SIDEBAR 12-2

Recommended Decontamination Equipment for Personnel

- Drop cloths of plastic or other suitable materials on which heavily contaminated equipment and outer protective clothing may be deposited
- Collection containers, such as drums or suitably lined trash cans, for storing disposable clothing and heavily contaminated PPE that must be discarded
- Lined box with absorbents for wiping or rinsing off gross contaminants and liquid contaminants
- Large galvanized tubs, stock tanks, or wading pools to hold wash and rinse solutions; these should be at least large enough for a worker to place a booted foot in and should have either no drain or a drain connected to a collection tank or appropriate treatment system
- Wash solutions selected to wash off and reduce the hazards associated with the contaminants
- Rinse solutions selected to remove contaminants and contaminated wash solutions
- Long-handled, soft-bristled brushes to help wash and rinse contaminated objects
- Paper or cloth towels for drying PPE
- Lockers and cabinets for storage of decontaminated clothing and equipment
- Metal or plastic cans or drums for contaminated wash and rinse solutions
- Plastic sheeting, sealed pads with drains, or other appropriate methods for containing and collecting contaminated wash and rinse solutions spilled during decontamination
- Shower facilities for full body wash or, at a minimum, personal wash sinks (with drains connected to a collection tank or appropriate treatment system)
- Soap or wash solution, single-use washcloths, and single-use towels for personnel
- Lockers or closets for clean clothing and personal item storage

Source: Adapted from NIOSH 1998.

- 3 gallons of water at the start of the workday or access to gently running water at the work site
- soap (cannot use only moist towelettes or hand sanitizer)
- single-use towels
- a clean pair of coveralls

Decontamination Equipment Selection

Sidebar 12-2 lists recommended equipment for decontamination of personnel and PPE. In selecting decontamination equipment, consider whether the equipment itself can be decontaminated for reuse or can be easily disposed of. Note that other types of equipment not listed in Sidebar 12-2 may be appropriate in certain situations.

Pesticide Leaks and Spills

Treat all pesticide leaks or spills as emergencies. Concentrated pesticide spills are much more dangerous than pesticides diluted with water, but both types should be dealt with immediately. For information on reporting requirements, visit the California Office of Emergency Services website, caloes.ca.gov.

Spill Kits

List the contents of a pesticide spill kit.

Keep a spill cleanup kit readily available whenever you handle pesticides or their containers. Also maintain a spill kit at the location where pesticides are mixed, loaded, and stored and on each vehicle that transports pesticides. If a spill occurs, you will not have the time or the opportunity to find all of the items needed to respond to the situation. Include the following in a kit:

- telephone numbers for emergency assistance
- gloves, footwear, and aprons that comply with both the label and SDS PPE requirements
- protective eyewear as required by the label and SDS
- an appropriate respirator if either the pesticide label or SDS require the use of one
- containment tubes or pads to confine the leak or spill

- absorbent materials such as spill pillows, absorbent clay, sawdust, cat litter, activated charcoal, vermiculite, or paper for liquid spills
- sweeping compound for dry spills
- a shovel, broom, and dustpan
- heavy-duty detergent
- a fire extinguisher rated for all types of fires
- any other spill cleanup items specified on the labeling of any products you use regularly
- a sturdy plastic container that holds the quantity of pesticide inside the largest pesticide container being handled and that can be tightly closed

Store spill kit items in a plastic container, replace items that have been used or discarded, and keep the contents clean and in working order until needed.

What to Do When Leaks and Spills Occur

Describe what to do when faced with a pesticide leak or spill.

Immediate Actions

- If the spill occurs on a public roadway, call 9-1-1 and the California Emergency Management Agency, 1-800-852-7550 (see "Reporting," below).
- If anyone has been injured or contaminated, call 9-1-1 and administer first aid until help arrives.
- Rope off the area or set up barricades to keep everyone away from the contaminated site.
- If the spill is indoors, get out of the building. If you have access to the proper PPE, reenter the building to open doors and windows and set up a portable fan.

Contaminated Materials

- Put materials that were contaminated by the spill or have been cleaned up into a sealable drum. Label the drum to indicate that it contains hazardous waste. Include the name of the pesticide and the signal word (DANGER, WARNING, or CAUTION). Because local regulations vary, contact the county agricultural commissioner or the Department of Toxic Substances Control regional office for instructions on how to dispose of the sealed drum and its contents. Under most circumstances, you must send the residue from a pesticide spill to a Class I disposal facility.
- Spills on cleanable surfaces such as concrete require thorough decontamination. Commercial decontamination preparations are available for this purpose, or you can prepare a solution that contains 4 tablespoons of detergent and 1 pound of soda ash dissolved in each gallon of water. Soda ash cannot be used for detoxification of certain pesticides, so check the label or SDS before using this solution. Contact the pesticide manufacturer if you have any questions. For more information, see the section "How to Deal with Pesticide Leaks or Spills," below.

Reporting

When spills occur on public roadways, you must

- call 9-1-1 (or the local emergency response agency)
- call 1-800-852-7550 (or 916-845-8911) and report the spill to the California Emergency Management Agency, California State Warning Center

If the spill causes injury or exposure, you must notify the California Occupational Safety and Health Administration (Cal/OSHA). If the spill occurs in or near a waterway, you must notify the United States Coast Guard; the Department of Fish and Wildlife, Office of Spill Prevention and Response; and the local Regional Water Quality Control Board. Report all leaks or spills of pesticides, no matter where they occur, to the local county agricultural commissioner as soon as possible. When you call the required agencies, you must provide

- your identity
- the location, date, and time of the spill, release, or threatened release

- the location of threatened or involved waterways or storm drains
- the substance and quantity involved
- the chemical name (if its toxicity level makes the substance extremely hazardous, report that information, as well)
- a description of what happened

If the spill exceeds federal reporting requirements, you will also need to report

- medium or media impacted by the release
- time and duration of the release
- proper precautions to take
- known or anticipated health risks
- name and phone number where officials can reach you if they need more information

In addition, a written report may be required. Check California laws and regulations to find out if a written report is required in your situation and how soon it must be delivered to avoid emergency notification penalties.

How to Deal with Pesticide Leaks or Spills

The types of pesticide leaks and spills you will most likely encounter will be controllable quantities, such as when a container is damaged or slips to the ground or when diluted pesticide leaks from application equipment. Proper and immediate response to even these small leaks and spills is necessary to minimize damage to human and environmental health. Follow these basic steps when cleaning up a pesticide leak or spill:

- **Clear the area.** Keep people and animals away from the contaminated area. Provide first aid if anyone has been injured or contaminated. Send for medical help if necessary.
- **Prevent fires.** Some liquid pesticides are flammable or are formulated in flammable carriers. Pesticide powders are potentially explosive, especially if a dust cloud forms in an enclosed area. Do not allow any smoking near a spill. If the spill occurs in an enclosed area, shut off all electrical appliances and motors that could produce sparks and ignite a fire or explosion. See "How to Deal with a Pesticide Fire," below.
- **Wear PPE.** Before beginning any cleanup, put on the PPE listed on the label for handling the concentrated material. Check the pesticide label for additional precautions. If you are uncertain what has been spilled, wear the maximum protection, which includes chemical-resistant boots and gloves, waterproof protective clothing, goggles, and a respirator.
- **Contain the leak.** Stop leaks by transferring the pesticide to another container or by patching the leaking container (repair paper bags and cardboard boxes with strong tape). Use soil, sand, sawdust, or absorbent clay to form a containment dam around liquid leaks. Common cat litter is a good absorbent material for pesticide cleanup. If the wind is blowing pesticide dusts or powders, cover the spill with a plastic tarp or, if a covering is not available, lightly spray the area with water to prevent offsite movement.
- **Clean up the pesticide.** Proceed to clean up the spill or leak (Fig. 12-4). Brush the containment dam of absorbent material toward the center of a liquid spill. Add additional

FIGURE 12-4.

Cover pesticide spills with an absorbent material and shovel it into a sealable container. When the cleanup is completed, seal and label the container and send it to a Class I disposal site. Wear personal protective equipment (required by the pesticide label) during the cleanup.

absorbent material if necessary. Sweep up granule formulations. If the spill is on soil, shovel out the top 2 to 3 inches of soil for disposal. Place the absorbent or spilled dry product and any contaminated soil in a sealable container. Containers for holding contaminated materials must be suitable for transporting. Label the container with the pesticide name and signal word.

- **Clean nonporous surfaces and safety equipment.** If the spill occurred on a cleanable surface such as concrete or asphalt, use a broom to scrub the contaminated surface with a strong detergent solution. Clean this up again with absorbent material and place it in the container. Equipment such as brooms, shovels, and dustpans must be cleaned or disposed of after use. For instance, brooms cannot be cleaned and so must be discarded, whereas a shovel can be appropriately decontaminated after use and can be kept. When you finish, clean or dispose of your PPE.
- **Dispose of the material.** Local regulations on disposal of hazardous materials may vary. Check with the local county agricultural commissioner or the California Department of Toxic Substances Control regional office for instructions on how to dispose of the container and its contents.

Other Types of Pesticide Emergencies

How to Deal with a Pesticide Fire

Describe what to do when faced with a pesticide fire.

If a pesticide fire breaks out:

- **Call the fire department.** Contact the nearest fire department as quickly as possible (call 9-1-1). Inform them that there is a fire involving pesticides. Provide them with the names of the chemicals contained in the structure or vehicle. If possible, provide Safety Data Sheets to the arriving fire units.
- **Clear the area.** Get people out of the immediate area of the fire; there may be considerable risk of toxic fumes and explosion.
- **Evacuate and isolate the area around and downwind of the fire.** Protect animals and move equipment and vehicles that could be damaged by the fire or fumes or that would impair firefighting efforts. Keep spectators from being exposed to smoke from the fire and runoff from firefighting. Contact the police or sheriff and have downwind residences, schools, and buildings evacuated until the danger has passed.

How to Deal with Stolen Pesticides

Describe what to do when a pesticide product has been stolen.

The first line of defense in any security program is properly trained employees and contractors. They notice much of what occurs in and around a pesticide storage facility or pesticide application business and can provide an early warning when something does not seem quite right or someone is acting suspiciously. Security training and awareness can ensure that these individuals can act effectively as an alert surveillance system. At a minimum, instruct all employees on pesticide inventory control, security of storage facilities and application equipment, and emergency preparedness and response, including notifying supervisors of the incident.

Notification Plan

Describe how to respond to the misapplication of pesticides.

If a breach of security or suspicious activity does occur, contact the appropriate authorities immediately. In addition to alerting the local police department and agricultural commissioner, you must immediately report any threats or suspicious behavior to the local FBI field office. These agencies also must be informed of incidents involving pesticide exposures that occur under circumstances inconsistent with a product's normal use pattern. Information on the location of the appropriate FBI office is available at the FBI website, fbi.gov.

Misapplication of Pesticides

Another form of emergency may exist when pesticides have been misapplied either intentionally, accidentally, or through negligence:

- Intentional misapplication involves intentional use of a pesticide on an unregistered site or crop or knowingly applying pesticides in a manner inconsistent with label directions.
- Accidental misapplication involves unknowingly applying a pesticide to a site not on the label.
- Negligent application involves improper calibration of application equipment as well as improper use and disposal of the pesticide; it also involves applying pesticides at the wrong time or in any other way inconsistent with label requirements.

Making an application mistake is a serious problem; do not compound the damage by failing to take responsible corrective action once the mistake is discovered. You may be financially responsible for damages, both physical and legal, caused by your misapplication of a pesticide; however, you may be able to reduce the amount of damage and liability by taking prompt action once you discover the error. Of primary importance is the protection of people, animals, and the environment.

Incorrect Amount of Pesticide Applied

Although using insufficient quantities of pesticides usually does not give adequate control of the target pest and is a waste of time and money, it generally presents no immediate problems to people or the environment. Using excessive amounts of pesticide, however, can be an environmental threat and a danger to human health, and it is illegal. This type of problem occurs as a result of

- poor calibration of your application equipment
- faulty mixing of chemicals in your spray tank
- not understanding the label statement regarding application rates

Residues from the pesticide may last longer than expected, or a concentrated application may cause damage to the treated area.

Correcting the Problem. Once an improper application has been discovered, take immediate action. Notify the local county agricultural commissioner of the problem and seek information and advice on what remedies to take. Contact the pesticide manufacturer to find out what corrective measures they suggest. Remember, speed is of the utmost importance when trying to reduce damage.

Wrong Pesticide Applied

Lack of attention to your mixing operation or giving the wrong instructions to an employee may result in the wrong pesticide being applied. Besides possible damage to plants or surfaces in the treatment area, using the wrong pesticide exposes you and your workers to unanticipated hazards. For instance, mixing and application might take place without the required PPE, resulting in possible injury to the applicator.

Correcting the Problem. When you discover that you have mixed or applied the wrong pesticide, contact the county agricultural commissioner for help, then call the pesticide manufacturer. Notify people in the application area, and keep them away until it can be made safe again.

Pesticides Applied to the Wrong Crop

Another form of accident involves pesticides being applied to the wrong crop. This can be a serious problem if the crop is not listed on the pesticide label or if there are workers in the field who are performing cultural operations.

Correcting the Problem. Contact your local county agricultural commissioner and the pesticide manufacturer for assistance. Notify people in the application area, and keep them away until it can be made safe again.

Reviewing Emergency Response to Accidents

Explain why any incident should be reviewed.

Accidents happen. The best way to be prepared for accidents is to review emergency responses to past incidents. Good recordkeeping, video or photographic evidence, and first-hand reporting from those involved in the incident can help you and your co-workers understand what went wrong and learn how to respond more effectively to future incidents. Questions to ask include

- What caused the accident?
- What pesticide(s) were involved in the accident?
- How did people respond as the situation unfolded?
- How could people improve their response to the situation?

Emergency response drills are another way to help you understand how to respond in case of an actual emergency. These drills can mimic pesticide fires or spills and will test your knowledge of your company's emergency response plan. Records of these drills can then be reviewed in order to find the strengths and weaknesses of your response plan. Aspects of the plan that were implemented well should be complimented, and areas of weakness can be addressed through additional, focused training.

Chapter 12 Review Questions

1. **The help you give people who are exposed to pesticides before they can be treated by a medical professional is called ______.**
   - ☐ a. practical treatment
   - ☐ b. first aid
   - ☐ c. emergency care

2. **Match the emergency type with the procedures to follow when providing first aid.**

1. pesticides on skin or clothing	a. Prevent chilling (from shock) by wrapping the person in a blanket after removing them from the accident site and disposing of contaminated clothing.
2. inhaled pesticides	b. After removing contaminated clothing, thoroughly wash the affected areas with soap or detergent and large amounts of water.
3. heat-related illness	c. Move the affected person to an air-conditioned or cool, shaded area.

3. **How can you tell if someone is suffering from acute pesticide poisoning?**
   - ☐ a. Find out what immediately preceded the onset of symptoms.
   - ☐ b. Look to see if there are any obvious signs of a pesticide spill nearby.
   - ☐ c. It is impossible to tell unless you are a trained medical professional.

4. **A spill kit should contain which of the following items? Select all that apply.**
   - ☐ a. shovel, broom, and dustpan
   - ☐ b. wash and rinse buckets
   - ☐ c. absorbent clay, sawdust, or cat litter
   - ☐ d. PPE as required by the pesticide label
   - ☐ e. large galvanized tubs

5. **Why is it important to review the response to a pesticide accident after it is over?**
   - ☐ a. Looking back at an incident gives everyone a chance to process what happened and move on in a positive way.
   - ☐ b. Thorough review of the response to an accident can help everyone respond more effectively to future incidents.
   - ☐ c. Reviewing past incidents reduces liability and insurance rates, since employees will be better prepared to respond to emergencies.

APPENDIX A

Definition of "pesticide" according to California regulation

Division 6. Pesticides and Pest Control Operations
Chapter 1. Pesticide Regulatory Program
Subchapter 1. Definition of Terms
Article 1. Definitions for Division 6
6000. Definitions.
"**Pesticide**" means:

(a) Any substance or mixture of substances that is a pesticide as defined in the Food and Agricultural Code and includes mixtures and dilutions of pesticides;

(b) As the term is used in Section 12995 of the Food and Agricultural Code, includes any substance or product that the user intends to be used for the pesticidal poison purposes specified in Sections 12753 and 12758 of the Food and Agricultural Code.

APPENDIX B

References to specific sections of the California Code of Regulations (CCR Title 3) around minimal standards for all employees working in any capacity on a farm

6400. Restricted Materials.
6406. Supervision Standards.
6412. Restricted Material Permit Requirements.
6445. Fumigation-Handling Activities.
6600. General Standards of Care.
6602. Availability of Labeling.
6604. Accurate Measurement.
6606. Uniform Mixture.
6608. Equipment Cleaning.
6609. Wellhead Protection.
6610. Backflow Prevention.
6614. Protection of Persons, Animals, and Property.
6670. General Requirement.
6674. Posting of Pesticide Storage Areas.
6676. Container Requirements.
6678. Service Container Labeling.
6680. Prohibited Containers for Pesticides.
6682. Transportation.
6684. Rinse and Drain Procedures.
6686. Exemptions.
6700. Scope.
6702. Employer-Employee Responsibilities.
6720. Safety of Employed Persons.
6723. Hazard Communication for Pesticide Handlers.
6723.1. Application-Specific Information for Handlers.
6724. Handler Training.
6726. Emergency Medical Care.
6730. Working Alone.
6732. Change Area.
6734. Handler Decontamination Facilities.
6738. Personal Protective Equipment.
6739. Respiratory Protection.
6740. Adequate Light.
6742. Safe Equipment.
6744. Equipment Maintenance.
6746. Closed Systems.
6760. Employer Responsibility and Exceptions.
6761. Hazard Communication for Field Workers.
6761. Application-Specific Information for Fieldworkers.
6762. Field Work during Pesticide Application.
6764. Fieldworker Training.
6766. Emergency Medical Care.
6768. Fieldworker Decontamination Facilities.
6770. Field Entry after Scheduled or Completed Pesticide Application.
6771. Requirements for Early-Entry Employees.
6772. Restricted-Entry Intervals.
6776. Field Postings.
6780. General Fumigation Safe-Use Requirements.
6782. Fumigation of Enclosed Spaces.
6784. Field Fumigation.
6990. Pesticide Use Near School Sites.
6991. Pesticide Application Restrictions.
6992. Annual Notification.

APPENDIX C

Forms needed to document safety training

Fieldworker Training Record 190
Handler Safety Training Record 192
Employer's Written Handler Training Program 193

Pesticide Training

Fieldworker Training Record

Name of Employer:____________________________ Date of training: _______________

Crew Identification (optional): __________________ Crew foreman (optional): ________________

See Next Page for Name(s) of Trainees

Name of Trainer: ______________________

☐ Certified Applicator: Type _______________
Lic/Cert # _______________

☐ UCCE Advisor

☐ Instructor Training Program (attach certificate copy)

☐ Pest Control Adviser License # _______________

☐ Registered Professional Forester # ____________

☐ County Biologist License from CDFA

☐ Other DPR-approved qualification (attach copy)

<u>Fieldworker Pesticide Safety Training Topics</u> required by 3CCR section 6764:

- ☐ - Where and in what forms pesticides and residues may be encountered
- ☐ - Potential hazards that pesticides present to fieldworkers and their families, including acute, chronic, delayed, and sensitization effects
- ☐ - Routes pesticides can enter the body
- ☐ - Signs and symptoms of overexposure
- ☐ - Prevention, recognition, and first aid for heat-related illnesses per 8CCR section 3395
- ☐ - Wear work clothing that protects the body from pesticide residues. Wash work clothes separately from other laundry.
- ☐ - Information provided by Safety Data Sheets
- ☐ - Hazard communication requirements (Pesticide use records, Application-Specific Information, SDS availability, Completed A-9 location)
- ☐ - Routine decontamination procedures and the employer's responsibility to provide decontamination supplies
  - Wash hands before eating, drinking, using the toilet, chewing gum, or using tobacco;
  - Thoroughly wash or shower with soap and water;
  - Change into clean clothes as soon as possible.
- ☐ - First aid and emergency decontamination procedures (including emergency eye flushing and procedures if pesticides are spilled or sprayed on the employee)
- ☐ - How and when to obtain emergency medical care
- ☐ - REIs and what field posting means, including both California and federal field posting signs
- ☐ - Keep out of application exclusion zones
- ☐ - Employees must be at least 18 years old to perform early-entry activities. Employees must be provided specific information before directed to perform early-entry activities
- ☐ - Employees cannot be allowed or directed to handle pesticides unless the employee has been trained as a handler
- ☐ - Employees must not take pesticides home
- ☐ - Potential hazards to children and pregnant women from pesticide exposures:
  - Children and nonworking family members should keep away from pesticide-treated fields;
  - After work, remove boots or shoes before entering the home and remove work clothes;
  - Wash or shower before physical contact with children or family members.
- ☐ - How to report suspected pesticide use violations
- ☐ - Employee rights, including the right:
  - To receive information about pesticides
  - For the employee's physician or employee representative designated in writing to receive information
  - To be protected against retaliation
  - To report suspected violations to DPR or county agricultural commissioner.

- Each employee must be trained within the last 12 months and before assignment to work in a treated field
- Training shall be presented in a manner the employee can understand using nontechnical terms, including response to questions
- Training location reasonably free from distraction and trainer must be present throughout entire presentation
- This record must be kept for two years, must be accessible to the employee, and must be provided to the employee, the County Agricultural Commissioner or the Department of Pesticide Regulation upon request

____

Continued on next page

(Rev. 6/2018)

Pesticide Training

Fieldworker Training Record (continued)

Name of Employer:_____________________ Date of training: __________

Crew Identification (optional): ____________ Crew foreman (optional): __________

Training Materials Used (Include videos, pamphlets, PSIS, or other training materials):

Title	Source

Use additional pages if necessary

	Print Your Name	**Sign Your Name**	**Employee ID# (optional)**
1			
2			
3			
4			
5			
6			
7			
8			
9			
10			
11			
12			
13			
14			
15			
16			
17			
18			
19			
20			

(Rev. 6/2018) Page ____ of ____

Pesticide Training

Handler Safety Training Record

Pursuant to 3 CCR section 6724

Training is in accordance with Employer's Written Handler Training Program

Print EMPLOYER's name: ______________________ Initial/Annual Training Date: ______________

Print EMPLOYEE's name*: ______________________ Print TRAINER's name: ______________

EMPLOYEE's signature: ______________________ Trainer Qualification*: ______________

ASSIGNED JOB DUTIES Trainer Lic/Cert #*: ______________

☐ Mixer/Loader ☐ Service/Repair

☐ Applicator ☐ Flagger ☐ Other: ________

Title(s) and source(s) of the training materials used*:

____

____

____

* Required for employee pesticide training for the production of agricultural commodities.

Pesticide (Attach additional pages if necessary)	READ THE LABEL: Signal word, precautionary statements, PPE, first aid, rate, dilution volume	SAFETY REQUIRE-MENTS and procedures, including engineering controls (such as closed mixing systems and enclosed cabs)	HAZARDS OF THE PESTICIDE including acute, chronic, and delayed effects, and sensitization effects from labeling, SDS, or other sources	SIGNS AND SYMPTOMS of overexposure	Trainer Initials	Employee Initials	Date Employee Trained on Pesticide

The employer must keep this record for two years at a central location at the workplace accessible to employees.

Pesticide Training

Employer's Written Handler Training Program

Pursuant to 3 CCR section 6724

____

(Name of Employer)

Name of Firm or Person Providing Training:

____

Note: For the commercial or research production of an agricultural plant commodity, the employer must assure the trainer is one of the following at the time of training (trainer licenses or certificates must be valid):

- A California certified applicator (commercial or private) or Agricultural Pest Control Adviser
- A County Biologist License in Pesticide Regulation or Investigation and Environmental Monitoring
- A University of California Extension Advisor
- A person who has completed nan "instructor training" program
- A California Registered Professional Forester
- Other trainer qualification approved by DPR

Training Materials:

Name of videos, pamphlets, study guides, or other training materials, and a brief description

1 ___________________________________

2 ___________________________________

3 ___________________________________

4 ___________________________________

5 ___________________________________

(Attach additional pages if necessary)

Circle Pesticide Safety Information Series (PSIS) Leaflets Used:

A-1 A-2 A-3 A-4 A-5 A-6 A-7 A-8 A-9 A-10

N-1 N-2 N-3 N-4 N-5 N-6 N-7 N-8

Pesticide product labeling and Safety Data Sheets (SDSs) as noted below:

	Product Name	Labeling?	SDS?
1		Yes / No	Yes / No
2		Yes / No	Yes / No
3		Yes / No	Yes / No
4		Yes / No	Yes / No
5		Yes / No	Yes / No
6		Yes / No	Yes / No
7		Yes / No	Yes / No
8		Yes / No	Yes / No
9		Yes / No	Yes / No
10		Yes / No	Yes / No
11		Yes / No	Yes / No

(Attach additional pages if necessary) (Circle Yes or No)

Continued on next page

Pesticide Training

Employer's Written Handler Training Program (continued)

Pursuant to 3 CCR section 6724

____

(Name of Employer)

This WRITTEN PROGRAM went into effect on (date): ______________________

This WRITTEN PROGRAM was retired on (date): ______________________

- A copy of the training program must be maintained while in use and for two years after use, at a central location at the workplace.
- Employee training shall be in a manner the employee can understand, be conducted pursuant to this written training program, and include response to questions.
- Training shall be completed before the employee is allowed to handle pesticides, continually updated to cover any new pesticides that will be handled or training topics to be covered, and repeated at least annually thereafter.
- Training for employees handling pesticides used for the commercial or research production of an agricultural commodity must be at a location reasonably free from distraction and trainers must be present throughout the entire presentation.

The following topics are applicable to my employees handling pesticides and are covered in initial and annual training for each pesticide or chemically-similar group of pesticides handled:

Circle if applicable	Training Topic
Yes No	Format and meaning of information, such as precautionary statements about human health hazards, contained in pesticide product labeling
Yes No	Applicator's responsibility to protect persons, animals, and property while applying pesticides; and not to apply pesticides in a manner that results in contact with persons not involved in the application process
Yes No	Need for, limitations, appropriate use, removal, and sanitation of any required personal protective equipment
Yes No	Safety requirements and procedures, including engineering controls (such as closed mixing systems and enclosed cabs) for handling, transporting, storing, disposing of pesticides, and spill clean-up
Yes No	Where and in what forms pesticides may be encountered, including treated surfaces, residues on clothing, personal protective equipment, application equipment, and drift
Yes No	Hazards of pesticides, including acute, chronic, and delayed effects, and sensitization effects, as identified in pesticide product labeling, Safety Data Sheets, or Pesticide Safety Information Series leaflets
Yes No	Routes by which pesticides can enter the body
Yes No	Signs and symptoms of overexposure
Yes No	Routine decontamination procedures when handling pesticides, including that employees should: A. Wash hands before eating, drinking, using the toilet, chewing gum, or using tobacco; B. Thoroughly wash or shower with soap and water; C. Change into clean clothes as soon as possible; and D. Wash work clothes separately from other laundry before wearing them again.

Continued on next page

Pesticide Training

Employer's Written Handler Training Program (continued)

Pursuant to 3 CCR section 6724

____

(Name of Employer)

Circle if applicable	**Training Topic**
Yes No	How Safety Data Sheets provide hazard, emergency medical treatment, and other information about the pesticides with which employees may come in contact
Yes No	The hazard communication program requirements of section 6723
Yes No	The purposes and requirements for medical supervision if organophosphate or carbamate pesticides with the signal word "DANGER" or "WARNING" on the labeling are mixed, loaded, or applied for the commercial or research production of an agricultural plant commodity
Yes No	First aid and emergency decontamination procedures and emergency eye flushing techniques; and if pesticides are spilled or sprayed on the body to wash immediately with decontamination supplies and as soon as possible, wash or shower with soap and water and change into clean clothes
Yes No	How and when to obtain emergency medical care
Yes No	Prevention, recognition, and first aid for heat-related illness in accordance with Title 8 of the California Code of Regulations, section 3395
Yes No	Requirements of Title 3, California Code of Regulations, Division 6, Chapters 3 and 4 relating to pesticide safety, Safety Data Sheets, and Pesticide Safety Information Series leaflets
Yes No	The requirement that handlers of pesticides used in the commercial or research production of an agricultural commodity must be at least 18 years of age
Yes No	Environmental concerns such as drift, runoff, and wildlife hazards
Yes No	Field posting requirements and restricted entry intervals when pesticides are applied for the commercial or research production of an agricultural commodity
Yes No	That employees should not take pesticides or pesticide containers home from work
Yes No	Potential hazards to children and pregnant women from pesticide exposures, including: A. Keeping children and nonworking family members away from treated areas; B. After handling pesticides, employees should remove boots or shoes before entering the home and remove work clothes; and C. Employees should wash or shower before physical contact with children or family members.
Yes No	How to report suspected pesticide use violations
Yes No	The employee's rights, including the right: A. To personally receive information about pesticides to which he or she may be exposed; B. For his or her physician or an employee representative designated in writing to receive information about pesticides to which he or she may be exposed; C. To be protected against retaliatory action due to the exercise of any of his or her rights; and D. To report suspected use violations to the Department or county agricultural commissioner.

APPENDIX D

California Code of Regulations Title 3 decontamination regulations

SECTION 6734. HANDLER DECONTAMINATION FACILITIES

(a) The employer shall assure that sufficient water, soap, and single use towels for routine washing of hands and face and for emergency eye flushing and washing of the entire body are available for employees as specified in this section.

  (1) This water shall be of a quality and temperature that will not cause illness or injury when it contacts the skin or eyes or if it is swallowed, and shall be stored separate from that used for mixing with pesticides unless the tank holding water for mixing with pesticides is equipped with appropriate valves to prevent back flow of pesticides into the water.

  (2) One clean change of coveralls shall be available at each decontamination site.

(b) For employees handling pesticides used in the commercial or research production of an agricultural commodity, the employer shall assure:

  (1) The water required to be available in (a) is at least 3 gallons per handler at the beginning of each handler's work day.

  (2) Hand sanitizing gels and liquids or wet towelettes are not used to meet the requirement for soap and single use towels as specified in (a).

  (3) The decontamination site is at the mixing/loading site and not more than ¼ mile (or at the nearest point of vehicular access) from other handlers, except that the decontamination site for pilots may be at the loading site regardless of distance from where the pilot is working. The decontamination site must not be in an area being treated or under a restricted-entry interval unless:

    (a) The handlers for whom the site is provided are working in that area being treated or under a restricted-entry interval;

    (b) The soap, towels, and extra change of coveralls are in an enclosed container; and

    (c) The water is running tap water or enclosed in a container.

  (4) Employees are notified of the location of the decontamination site prior to handling pesticides.

  (5) One pint of water for emergency eye flushing is immediately available to each employee (carried by the handler or on the vehicle or aircraft the handler is using) if the pesticide product labeling requires protective eyewear. When the handler is mixing or loading a pesticide, then only the requirements in (6) apply.

  (6) At the mixing/loading site there is immediate employee access to at least one system capable of delivering gently running water at a rate of least 0.4 gallons per minute for at least 15 minutes, or at least 6 gallons of water in containers suitable for providing a gentle eyeflush for about 15 minutes for emergency eye-flushing, if the product labeling requires protective eyewear or a closed mixing system is used.

(c) The decontamination site for employees handling pesticides for uses other than the commercial or research production of an agricultural plant commodity shall be within 100 feet of the mixing/loading site when they are handling pesticides with the signal word "DANGER" or "WARNING" on the label.

NOTE: Authority cited: Section 12981, Food and Agricultural Code.
Reference: Sections 12980 and 12981, Food and Agricultural Code.

SECTION 6768. FIELDWORKER DECONTAMINATION FACILITIES

(a) The employer shall assure that sufficient water and the following are located together at the decontamination site and reasonably accessible for washing of hands and face and for emergency eye flushing to all fieldworkers engaged in activities involving contact with treated surfaces in treated fields:

(1) At least 1 gallon of water per employee, or 3 gallons of water per employee for employees engaged in early-entry activities pursuant to section 6770(d). The water must be provided at the start of the work day and be of a quality and temperature that will not cause illness or injury when it contacts the skin or eyes or if it is swallowed. The water shall be stored separate from that used for mixing with pesticides unless the tank holding water for mixing with pesticides is equipped with appropriate valves to prevent back flow of pesticides into the water;

(2) Soap (hand sanitizing gels and liquids or wet towelettes do not meet the requirement for soap); and

(3) Single use towels (wet towelettes do not meet the requirement for single-use towels).

(b) The decontamination facilities shall be not more than ¼ mile from the fieldworkers (or at the nearest point of vehicular access). Employees must be notified of the location of the decontamination site prior to working in a treated field.

(c) The decontamination facilities shall not be in an area under a restricted-entry interval unless the fieldworkers for whom the site is provided are performing early-entry activities. The facilities shall not be in an area under treatment.

NOTE: Authority cited: Section 12981, Food and Agricultural Code. Reference: Sections 12980 and 12981, Food and Agricultural Code.

Answers to Review Questions

Chapter 1

1. b
2. c
3. b, c, e, f
4. a
5. c
6. a

Chapter 2

1. c
2. 1.a, 2.b, 3.b, 4.a
3. c
4. 1.c,h; 2.a,f; 3.b,e; 4.d,g
5. b
6. a
7. a
8. b

Chapter 3

1. a
2. 1.d,f; 2.b,e; 3.a,g; 4.c,h
3. 1.b, 2.c, 3.a
4. c
5. b
6. b
7. c
8. 1.b, 2.d, 3.a, 4.e, 5.c

Chapter 4

1. a
2. b
3. 1.c,e; 2.a; 3.b,f; 4.d
4. 1.b, 2.a, 3.a, 4.b
5. a.F, b.T, c.T, d.F, e.T, f.T

Chapter 5

1. a.T, b.F, c.T, d.F, e.T
2. 1.c, 2.d, 3.a, 4.b
3. 1.a, 2.c, 3.b, 4.a, 5.b, 6.c
4. a
5. b
6. a

Chapter 6

1. b
2. a, b, c, d
3. b, c
4. c
5. a, b
6. c
7. a

Chapter 7

1. a
2. c
3. b
4. 1.a, 2d, 3.b, 4.f, 5.g, 6.e, 7.c
5. 1.c, 2.a, 3.b
6. a.T, b.T, c.F, d.F, e.T
7. a

Chapter 8

1. a
2. a
3. c
4. b
5. 1.b, 2.c, 3.a
6. b
7. c

Chapter 9

1. 1.c, 2.b, 3.e, 4.a, 5.d
2. a.F, b.T, c.T, d.F, e.T, f.F, g.T, h.T
3. 1.b, 2.c, 3.d, 4.a, 5.e
4. b, d, e
5. 1.c, 2.d, 3.a, 4.b

Chapter 10

1. c
2. b, c, e, g, h
3. a
4. c
5. b
6. c

Chapter 11

1. b
2. b
3. a
4. e, a, d, c, b
5. 1.b, 2.a, 3.d, 4.e, 5.c
6. a.F, b.F, c.T, d.T, e.F
7. b

Chapter 12

1. b
2. 1.b, 2.a, 3.c
3. a
4. a, c, d
5. b

Glossary

abiotic. Nonliving factors such as wind, water, temperature, and soil type or texture.

abiotic disorders. Noninfectious diseases introduced by adverse environmental conditions, often as a result of human activity.

absorb. To soak up or take in a liquid or powder.

acaricide. A pesticide used to control mites.

accidental misapplication. An unintentional, incorrect application of a pesticide.

accumulate. To increase in quantity within an area, such as in the soil or tissues of a plant or animal.

acidic. A solution or substance that has a pH lower than 7.0.

activator. An adjuvant that increases the activity of a pesticide by reducing surface tension or speeding up penetration through insect or plant cuticles.

active ingredient (a.i.). The material in the pesticide formulation that actually destroys the target pest or performs the desired function.

acute effect. A symptom that becomes apparent soon after an exposure to a pesticide occurs.

acute onset. Symptoms of pesticide-related injury that appear soon after the exposure incident.

adjuvant. A material added to a pesticide mixture to improve or alter the deposition, toxic effects, mixing ability, persistence, or other qualities of the active ingredient.

adsorb. To gather and hold on a surface, such as pesticides that become attached to soil particles.

agitator. A mechanical or hydraulic device that stirs the liquid in a spray tank to prevent the mixture from separating or settling.

agricultural commissioner. The official in each county in California who has the responsibility for enforcing the state and federal pesticide regulations and issuing permits for restricted-use pesticides.

agricultural use. A classification of certain pesticides that limits their use to production agriculture settings.

a.i. See ***active ingredient***.

air blast sprayer. A sprayer that uses a high-powered fan to carry spray droplets to target surfaces; air blast sprayers are usually used on tall plants such as trees or vines.

air gap. A space between the filling hose and the liquid in the pesticide tank that prevents backflow of pesticide liquids into the water source.

alkaline. A solution or substance that has a pH greater than 7.0.

alternate hosts. Plants that support the survival of a pest when its main host is not available.

amphibian. Cold-blooded organism such as a frog, toad, or salamander.

annual. A plant that passes through its entire life cycle in 1 year or less. Plants can be further divided into summer or winter annuals.

apiary. A place where bees are kept, such as a beehive.

application pattern. The course the applicator follows through the area being treated with a pesticide.

application rate. The amount of pesticide that is applied to a known area, such as an acre.

application swath. See ***swath*** and ***swath width***.

aquatic. Pertaining to water, such as aquatic weeds or aquatic pest control.

aquifer. An underground formation of sand, gravel, or porous rock that contains water; the place where groundwater is found.

arthropod. An animal with jointed appendages and an external skeleton, such as an insect, spider, mite, crab, or centipede.

artificial respiration. See ***rescue breathing***.

atmosphere. The air or climate in a given place.

atmosphere-monitoring device. A piece of equipment used to detect and measure vapor levels in an enclosed area. Typically used after fumigation to ensure an area is safe to enter.

attractant. A substance that attracts a specific species of animal to it. When manufactured to attract pests to traps or poisoned bait, attractants are considered to be pesticides.

auger. A spiral-shaped shaft used for moving pesticide dusts or granules from a hopper to a moving belt or disc for application.

augmentation. The process of building up a population of natural enemies in an area by bringing in additional eggs, larvae, or adults of that species.

auricle. A small earlike projection from the base of a leaf or petal.

avicide. A pesticide used to control pest birds.

backflow. See ***back siphoning***.

backpack sprayer. Also known as a knapsack sprayer, a small, portable sprayer carried on the back of the person making the pesticide application; some are hand-operated, and others are powered by small gasoline engines.

back siphoning. The process that permits pesticide-contaminated water to be sucked from a spray tank back into a well or other water source.

bacteria. Unicellular, microscopic, plantlike organisms that live in soil, water, organic matter, or the bodies of plants and animals. Some bacteria cause plant or animal diseases.

bait. Food or foodlike substance that is used to attract and often poison pest animals.

bait station. A box or similar device designed to hold poisoned bait for controlling rodents, insects, or other pests; usually with baffles or small openings to prevent access to the bait except by the target pest.

band application. The application of liquid or dry pesticides in bands or strips, usually to the soil, rather than over the entire area.

beneficial. Being helpful in some way to people, such as a beneficial plant or insect.

beneficial organisms. Living things that prey upon, attack, or parasitize pests or serve as pollinators.

biennial. A plant that completes part of its life cycle in 1 year and the remainder of its life cycle the following year.

bioaccumulation. The gradual buildup of certain pesticides in the tissues of living organisms after feeding on lower organisms containing smaller amounts of these pesticides.

biological control. The action of parasites, predators, pathogens, or competitors in maintaining another organism's numerical density at a lower average than would occur in their absence; may occur naturally in the field or be the result of manipulation or introduction of biological control agents by people.

biological factors. Life cycles, life stages, physical attributes, and other factors that protect certain organisms from the toxic effects of pesticides.

biology. The body structure, behavior, and other qualities of a particular organism or class of organisms.

biotype. Any population within a species that has a distinct genetic variation from other populations.

boom. A structure mounted on a truck, tractor, or other vehicle, or held by hand, to which spray nozzles are attached.

boom applicator. A pesticide application device with multiple nozzles spaced along a boom, making it possible to spray a wide swath; usually used for applying herbicides or other pesticides in field and row crops.

brand name. The registered or trade name given to a pesticide by its manufacturer or formulator.

breakdown. The process by which chemicals, such as pesticides, decompose into other chemicals.

broadcast application. A method of applying pesticides by dispersing them over a wide area.

broadleaves. One of the major plant groups, known as dicots, with net-veined leaves usually broader than grasses, and whose seedlings have two seed leaves (cotyledons).

broad-spectrum pesticide. A pesticide that is capable of controlling many different species or types of pests. Also known as nonselective pesticide.

buffer. An adjuvant that lowers the pH of a spray solution and, depending on its concentration, can maintain the pH within a narrow range even if acidic or alkaline materials are added to the solution.

buffer zone. Areas of a field, usually a minimum of one swath width, left unsprayed to protect nearby structures or sensitive areas from drift; also known as buffer strips.

caking. The process by which pesticide dusts pack and clump together, preventing proper application.

calibration. The process used to measure the output of pesticide equipment so that the proper amount of pesticide can be applied to a given area.

California Department of Food and Agriculture (CDFA). The state agency responsible for protecting and promoting agriculture in California.

California Department of Pesticide Regulation (DPR). The state agency responsible for regulating the use of pesticides in California.

carbamate. A class of pesticides commonly used for control of insects, mites, fungi, and weeds. N-methyl carbamate insecticides, miticides, and nematicides are cholinesterase inhibitors; subgroups include dithiocarbamates and thiocarbamates.

carcinogen. A cancer-causing substance or agent.

cardiopulmonary resuscitation (CPR). A procedure designed to maintain circulatory action after breathing and heartbeat has stopped.

carrier. The liquid or powdered substance that is combined with the active ingredient in a pesticide formulation; may also apply to the water or oil that a pesticide is mixed with prior to application.

CAUTION. Signal word used on labels of the least-toxic pesticides; pesticides with an oral LD_{50} greater than 500 and a dermal LD_{50} greater than 2,000.

certified pesticide applicator. A person who has demonstrated through an examination process the ability to safely handle and apply highly hazardous restricted-use pesticides.

certified private applicator. A property owner or manager, or a responsible person employed by the property owner or manager, who has demonstrated through an examination process the ability to safely handle and apply restricted-use pesticides on the property under their control.

chemical control. The use of naturally occurring or synthetic pesticides to manage pest populations in an area.

chemical family. A group of chemicals that have common characteristics, such as chemical structure or environmental persistence.

chemical injection system. The part of a chemigation system that controls the amount of pesticide injected into irrigation water.

chemical name. The official name given to a chemical compound to distinguish it from other chemical compounds.

chemigation. The application of pesticides to target areas through an irrigation system.

CHEMTREC. A chemical industry–supported organization that provides assistance and advice on pesticide emergencies; telephone 1-800-424-9300.

chronic. Of long duration or frequent recurrence.

chronic effect. The harmful effects that occur from small, repeated doses of pesticides over time.

chronic symptoms. Symptoms of pesticide poisoning that occur days, weeks, or months after the actual exposure.

Class I disposal site. A disposal site for toxic and hazardous materials such as pesticides and pesticide-contaminated wastes.

Class II disposal site. A disposal site for nontoxic and nonhazardous materials such as household and commercial waste.

classical biological control. A pest control method that uses natural enemies and is directed toward pests that are not native to a geographical area; it involves locating the native home of an introduced pest and finding suitable natural enemies that can be imported, reared, and released into the area where the pest has become established.

closed mixing system. A device used for measuring and transferring liquid pesticides from the original container to the spray tank to reduce the chances of exposure to concentrated pesticides; special packaging, such as water-soluble bags, is also considered a simple closed mixing system.

combine. To mix or unite.

common name. The recognized, nonscientific name given to living organisms; also, names of pesticides separate from their brand (trade) names and chemical names.

compatible. The condition in which two or more pesticides mix without unsatisfactory chemical or physical changes.

confined area. Places such as buildings or greenhouses, attics, crawl spaces, or holds of ships that may have restricted air circulation and therefore promote the buildup of toxic fumes or vapors from a pesticide application.

contact poison. A pesticide that provides control when target pests come in physical contact with it.

continuing education (CE). Approved classes, seminars, or trainings that certified or licensed applicators must take to keep their credentials valid. Topics include pesticide use and safety, California laws and regulations, and pest management.

convulsions. Contortions of the body caused by violent, involuntary muscular contractions; a possible symptom of pesticide poisoning.

corrosive materials. Certain chemicals that react with metals or other materials.

cotyledon. The first leaf or pair of leaves of a sprouted seed; grasses (monocots) have one cotyledon, while broadleaved plants (dicots) have two.

county agricultural commissioner. The official in each county in California who has the responsibility for enforcing the state and federal pesticide regulations and issuing permits for restricted-use pesticides; county agricultural commissioners and their staff frequently inspect pesticide applications and application sites and investigate pesticide illnesses and environmental exposures; all agricultural uses of pesticides must be reported monthly to county agricultural commissioners.

coverage. The degree to which a pesticide is distributed over a target surface.

coverall. A one- or two-piece garment of closely woven fabric that covers the entire body except the head, hands, and feet, and must be provided by the employer as personal protective equipment. Coveralls differ from, and should not be confused with, work clothing that can be required to be provided by the employee.

CPR. See ***cardiopulmonary resuscitation***.

cross-resistance. A condition in which an organism that has developed resistance to one type or group of pesticides is also resistant to other similar or dissimilar pesticides, even though the organism has never been exposed to those pesticides.

cultural controls. The modification of normal crop or landscape management practices to decrease pest establishment, reproduction, dispersal, survival, or damage.

cuticle. The outer protective covering of plants and arthropods that aids in preventing moisture loss.

DANGER. The signal word used on labels of highly hazardous pesticides that have an oral LD_{50} less than 50 or a dermal LD_{50} less than 200.

DANGER-POISON. The signal word used in combination with the skull and crossbones on labels of pesticides considered to be the most hazardous, having an oral LD_{50} less than 50 or a dermal LD_{50} less than 200; this signal word is also used to identify pesticides that can cause specific, serious health or environmental hazards.

decontamination. The process of removing or neutralizing contaminants that have accumulated on people, clothes, and equipment (for instance, thoroughly washing skin exposed to pesticides with soap and water).

deficient. Not having enough of a specific thing.

deficient oxygen condition. Condition where the oxygen concentration in air falls below 19%, making an area highly hazardous.

defoliation. The removal or stripping of leaves from a plant.

degrade. To break down or deteriorate chemically.

degree-day. The amount of heat that accumulates over a 24-hour period when the average temperature is 1 degree above the lower developmental threshold of an organism.

delayed effects. Illnesses or injuries that appear more than 24 hours after exposure to a pesticide.

deposit. To put something in a specific place.

deposition. The placement of pesticides on target surfaces.

deposition aid. An adjuvant that improves the ability of a pesticide spray to reach the target.

dermal. Pertaining to the skin; one of the major ways pesticides can enter the body.

dermatitis. Inflammation, itching, or irritation of the skin, as can be caused by pesticide exposure.

detoxification. The process of removing toxic substances or qualities.

dichotomous key. A series of sequentially paired statements that help to identify insects or other living organisms; a type of identification key.

dicot. Plants whose seedlings produce two leaves (cotyledons). Commonly called broadleaves.

diluent. The liquid or powdered material that is combined with the active ingredient during manufacture of a pesticide formulation; also, the water, petroleum oil, or other liquid mixed with the formulated pesticide before application.

directions for use. The instructions found on pesticide labels indicating the proper procedures for mixing and application.

disease. A condition caused by biotic or abiotic factors that impairs some or all of the normal functions of a living organism.

disease cycle. The stages of development of a pathogen and the effect of the disease, as it develops, on the host.

disorder. A functional abnormality (something that is not normal) or disturbance within an organism.

dispersal. The act of spreading pesticide droplets, dusts, or granules widely over a target area. Also, the spread of living organisms throughout the environment, such as fungal spores.

disposable. Designed to be thrown away after use.

disposal site. See ***Class I disposal site*** and ***Class II disposal site***.

dissolve. To pass into solution.

dormant. To become inactive, such as trees that become bare during winter.

dose. The measured quantity of pesticide.

DPR. See ***California Department of Pesticide Regulation***.

drift. The movement of pesticide particles, spray, or vapor through the air away from the application site.

dry flowable. A dry, granular pesticide formulation intended to be mixed with water for application.

dust. Finely ground pesticide particles, sometimes combined with other materials. Dusts are applied without mixing with water or other liquid.

early-entry worker. An employee who must enter a pesticide application site to perform cultural activities before the expiration of the restricted-entry interval.

economic damage. Damage caused by pests to plants, animals, or other items that results in loss of income or a reduction of value.

economic injury threshold. The point at which the value of the damage caused by a pest exceeds the cost of controlling the pest, therefore making it practical to use a control method.

ecosystem. The community of organisms in an area and their nonliving environment.

efficacy. The ability of a pesticide to produce a desired effect on a target organism.

electrostatic. An electrical charge that causes a pesticide liquid or dust to be attracted to the target surface.

emergence. The appearance of a plant through the surface of the soil.

emergency exemption from registration. A federal exemption from regular pesticide registration sometimes issued when an emergency pest situation arises for which no pesticide is registered that has a tolerance on the crop in question.

emulsifiable concentrate. A pesticide formulation consisting of a petroleum-based liquid and emulsifiers that enable it to be mixed with water for application.

emulsifier. An adjuvant added to a pesticide formulation to permit petroleum-based pesticides to mix with water.

emulsion. Droplets of petroleum-based liquids (oils) suspended in water.

encapsulation. A process by which tiny liquid droplets or dry particles are contained in polymer plastic capsules to slow their release into the environment and prolong their effectiveness.

enclosed cab. A compartment with an air filtering system that is installed on a tractor to protect the operator from pesticide exposure.

endangered species. Rare or unusual living organisms whose existence is threatened by human activity, including the use of some types of pesticides.

engineering controls. Devices that have been developed to protect people as they are mixing, loading, and applying pesticides in a variety of situations, such as enclosed cabs and closed mixing systems.

environment. All of the living organisms and nonliving features of a defined area.

environmental contamination. Spread of pesticides away from the application site into the environment, usually with the potential for causing harm to organisms.

EPA. See ***U.S. Environmental Protection Agency***.

eradicate. To destroy completely or put an end to (a pest).

establishment number. A number assigned to registered pesticides by the U.S. EPA that indicates the location of the manufacturing or formulation facilities of that product.

evaporate. The process of a liquid turning into a gas or vapor.

exclusion. A pest management technique that uses physical or chemical barriers to prevent certain pests from getting into a defined area.

exposure. The unwanted contact with pesticides or pesticide residues by people, other organisms, or the environment.

fallow. Cultivated land that is allowed to lie dormant during a growing season.

farm advisors. University of California specialists in most counties of California who serve as resources for residents of the state on pest management, water management, soil management, nutrition, and many other issues.

Federal Insecticide, Fungicide, and Rodenticide Act (FIFRA). The federal law that regulates pesticide registration, labeling, use, and disposal in the United States.

fencerow. The strip of soil under a fence.

fertilizer. An organic or synthetic substance usually added to or spread onto soil to increase its ability to support plant growth; sometimes mixed and applied with pesticides.

fibrous. A word used to describe thin, long, multibranching roots that form a dense clump.

field incompatibility. An incompatibility between pesticides mixed together in a spray tank that occurs during application; may result from changes in the temperature of the water used in the mix or changes in the length of time the spray mixture has been in the tank.

fieldworker. An employee of a farming operation who performs cultural practices on crops or agricultural soil.

fieldworker training. Specific training mandated by the U.S. EPA and the state of California to protect fieldworkers from pesticide hazards when they work in pesticide-treated areas.

filamentous. Long and threadlike.

first aid. The immediate assistance provided to someone who is injured, ill, or has been overexposed to a pesticide.

first aid statement. The section of a pesticide label that describes appropriate first aid needed by a person exposed to that pesticide.

fit check. The procedure that must be carried out each time a person puts on an organic vapor-filtering respirator and that involves (1) properly adjusting the straps; (2) closing the filters with the hands and inhaling to check for air leaks around the face seal; and (3) closing the exhalation valve and exhaling to check for air leaks through the filters. Also known as seal check.

fit test. A test that must be performed to check the proper fit of an organic vapor-filtering respirator each time a new respirator is issued.

flowable. Formulations that consist of finely ground particles of pesticide active ingredient mixed with a liquid, along with emulsifiers, to form a concentrated emulsion.

flow rate. The amount of pesticide being expelled by a pesticide sprayer or granule applicator per unit of time.

foliage. The leaves of plants.

forecast. A prediction of weather conditions for the near future.

formulation. A mixture of active ingredient(s) combined during manufacture with other materials added to improve the mixing and handling qualities of a pesticide.

frass. Solid fecal material produced by insects.

fruiting bodies. Special structures produced by fungi that contain the spores by which the organisms reproduce.

fumes. Smoke, gas, or vapor; the vapor phase of some pesticide active ingredients.

fumigant. Vapor or gas form of a pesticide used to penetrate porous surfaces for control of soil-dwelling pests or pests in enclosed areas.

fungi (sing., fungus). Multicellular lower plants lacking chlorophyll, such as a mold, mildew, rust, or smut; fungal bodies normally consist of filamentous strands called mycelium, and they reproduce through dispersal of spores.

gall. An abnormal swelling of plant tissue, which can be caused by insects, nematodes, and pathogens.

general-use pesticide. A pesticide that has been designated for use by the general public as well as by licensed or certified applicators and that usually has minimal hazards. It does not require a permit for purchase or use.

global positioning system (GPS). A worldwide navigation system that uses information received from orbiting satellites.

GPS. See ***global positioning system***.

granule. A dry formulation of pesticide active ingredient and other materials compressed into small, pebblelike shapes.

groundwater. Freshwater trapped in aquifers beneath the surface of the soil; one of the primary sources of water for drinking, irrigation, and manufacturing.

ground wheel-driven. A trailer-mounted dry or liquid pesticide applicator that gets the power to drive a pump, auger, or spinning disc from the movement of one of the trailer wheels as the unit is towed.

growth regulator. A chemical that disrupts the normal growth of an organism.

habitat. The place where plants or animals live.

habitat modification. Intentionally limiting the availability of one or more of a pest's survival requirements, making the environment less suitable for pest population growth.

half-life. The amount of time it takes for a pesticide to be reduced to half its original toxicity or effectiveness.

hand lens. A small magnifying glass used in monitoring for plant pests.

handler. A person who mixes, loads, transfers, applies (including chemigation), or assists with the application (including flagging) of pesticides; who maintains, services, repairs, cleans, or handles equipment used in these activities; who works with unsealed pesticide containers; who adjusts, repairs, or removes treatment site coverings; who incorporates pesticides into the soil; who enters a treated area during any application or before the REI has expired; or who performs crop advisor duties.

harvest interval. A period of time, as indicated by the pesticide label, that must elapse after a pesticide has been applied to an edible crop before the crop can be harvested legally.

hazard. Something that is potentially very dangerous.

Hazard Communication Program. Part of California's pesticide regulations that requires employers to provide information about pesticides and pesticide applications at the workplace.

hazardous materials. Pesticides that have been classified by regulatory agencies as being harmful to the environment or to people, require special handling, and must be stored and transported in accordance with regulatory mandates.

hazardous waste. A hazardous material for which there is no further use and which must be disposed of only through special hazardous material incineration or by transporting to a Class I disposal site.

heat illness. Potentially life-threatening overheating of the body under working conditions that lack proper preventive measures, such as drinking plenty of water, taking frequent breaks in the shade to cool down, and removing or loosening personal protective equipment during breaks.

herbaceous. A plant that is herblike, usually with little or no woody tissue.

herbicide. A pesticide used for the control of weeds.

herbivore. An animal that feeds on plants.

hibernate. Passing the winter in a resting or nonactive state.

honeydew. The sweet, sticky fluid secreted by plant-feeding insects such as aphids and scales.

host. A plant or animal species that provides sustenance for another organism.

host resistance. The ability of a host plant or animal to ward off or resist attack by pests or to be able to tolerate damage caused by pests.

hyphae (sing., hypha). Threadlike fibers that make up fungal mycelium.

identification key. A written and/or illustrated tool that provides a systematic way to identify and distinguish related living organisms.

impermeable. Having the ability to resist penetration by a substance or object.

impregnates. Items, such as flea collars, that have been manufactured with a certain pesticide in it.

incompatibility. A condition in which two or more pesticides are unable to mix properly or one of the materials chemically alters the other to reduce its effectiveness or produce undesirable effects on the target.

incompatible mixture. The result when two or more pesticides are combined and react to make the mixture unusable.

incorporate. To move a pesticide below the surface of the soil by discing, tilling, or irrigation; also, to combine one pesticide with another.

incubation period. The time between infection of a host by a pathogen and the appearance of disease symptoms.

inert. Not having any chemical activity.

inert ingredients. Obsolete term for ingredients other than the active ingredient in a pesticide formulation; see ***other ingredients***.

infection. The establishment of a microorganism within the tissues of a host plant or animal.

infestation. A troublesome invasion of pests within an area such as a building, greenhouse, agricultural crop, or landscaped location.

infiltration. The movement of water into the soil.

ingest. To take into the body through the mouth, such as eating or swallowing.

inhale. To take into the body through the nose or mouth via the lungs.

inherit. To receive a characteristic or quality as a result of its being passed on genetically.

inhibit. To prevent a biochemical reaction within the tissues of a plant or animal.

injury. The physical harm or damage done (to an organism).

inoculate. To introduce a pathogen into an organism.

insect. A small arthropod that has six legs and sometimes one or two pairs of wings.

insecticide. A pesticide used for the control of insects; some insecticides are also labeled for control of ticks, mites, spiders, and other arthropods.

integrated pest management (IPM). A pest management program that uses life history information and extensive monitoring to understand a pest and its potential for causing economic damage. Control is achieved through multiple approaches including prevention, cultural practices, pesticide applications, exclusion, natural enemies, and host resistance. The goal is to achieve long-term suppression of target pests with minimal impact on nontarget organisms and the environment.

intentional misapplication. The deliberate improper use of a pesticide, such as knowingly exceeding the label rate or applying the material to a site not listed on the label.

interval. The legal period of time between when a pesticide is applied and when workers are allowed to enter the treated area or produce can be harvested; see also ***preharvest interval*** and ***restricted-entry interval***.

inversion. Weather phenomenon in which cool air near the ground is trapped by a layer of warmer air above; also known as a temperature inversion or inversion layer.

invertebrate. Any animal not having an internal skeleton or shell, such as insects, spiders, mites, worms, nematodes, snails, and slugs.

invert emulsion. An emulsion in which water droplets are suspended in an oil rather than the oil droplets being suspended in water.

IPM. See ***integrated pest management***.

irreversible. Not able to be undone or altered, as when a pesticide causes irreversible damage that cannot be corrected or medically treated.

irrigation. A method of supplying land or crops with water.

key pest. A pest that regularly causes major damage in a crop or landscape unless it is controlled.

knapsack sprayer. See ***backpack sprayer***.

knowledge expectations. The breadth of knowledge about an occupation or procedure, such as pesticide handling, that a person performing this job is expected to have as established by regulations and tested by certification examinations.

labeling. The pesticide label and all associated materials, including supplemental labels, special local needs registration information, and manufacturer's information.

larvae (sing., larva). The active, immature form of insects that undergo complete metamorphosis to reach adulthood.

LC_{50}. The lethal concentration of a pesticide in the air or in the body or water that will kill half of a test animal population; values are given in micrograms per milliliter of air or water (µg/ml).

LD_{50}. The lethal dose of a pesticide that will kill half of a test animal population; values are given in milligrams per kilogram of test animal body weight (mg/kg).

leaching. The process by which some pesticides move down through the soil, usually by being dissolved in water, with the possibility of reaching groundwater.

legible. Clear enough to be read; easily readable.

lethal. Capable of causing death.

lethal concentration. See ***LC50***.

lethal dose. See ***LD50***.

liability. Legal responsibility for something, especially costs or damages.

liability insurance. An insurance policy that covers the cost of damages from accidents or the improper use of pesticides.

life stages. The development stages living organisms pass through over time.

ligule. A thin outgrowth or fringe of hairs occurring at the collar region in many grass species.

mechanical controls. Devices that exclude, trap, or destroy pests, or modify the environment to make an area unsuitable for pests.

medical facility. A clinic, hospital, or physician's office where immediate medical care for pesticide-related illness or injury can be obtained.

mesh. The number of wires per inch in a screen, such as a screen used to filter foreign particles out of spray solutions; also used to describe the size of pesticide granules, pellets, and dusts.

metabolism. The total chemical process that takes place in a living organism to use food and manage wastes, provide for growth and reproduction, and accomplish all other life functions.

metamorphosis. The changes that take place in certain types of living organisms, such as insects, as they develop from eggs through adults.

microencapsulated materials. A formulation in which particles of the active ingredient are encased in plastic capsules; pesticide is released after application when the capsules break down.

microorganism. An organism of microscopic size, such as a bacterium, virus, fungus, viroid, or mycoplasma.

mimic. To copy or appear to be like something else.

mitigation. The process of making a problem such as a pest infestation less severe.

mixing. The process of opening pesticide containers, weighing or measuring specified amounts, and transferring these materials into application equipment, all in accordance with instructions found on pesticide labels.

mobile. Able to move freely or easily.

mode of action. The way a pesticide reacts with a pest organism to destroy it.

molluscicide. A pesticide used to control slugs and snails.

molting. A process of shedding the outer body covering or exoskeleton in invertebrates such as insects and spiders. Molting usually takes place to allow the animal to grow larger.

monitoring. The process of carefully watching the activities, growth, and development of pest organisms over a period of time, often utilizing very specific procedures.

monocot. A member of a group of plants whose seedlings have a single cotyledon.

monthly pesticide use report. A form that must be completed and submitted to the local agricultural commissioner's office by the 10th of the month following any month in which pesticides are applied to an agricultural crop.

MSDS. See *Safety Data Sheet*.

mycelium (pl., mycelia). The vegetative body of a fungus, consisting of a mass of slender filaments called hyphae.

National Institute for Occupational Safety and Health (NIOSH). The federal agency that tests and certifies respiratory equipment for pesticide application.

native. Animals or plants that are indigenous to an area.

natural enemy. An organism that can kill a pest organism, including predators, pathogens, parasites, and competitors.

necrosis. Localized death of living tissue.

negligent application. A pesticide application in which the applicator fails to exercise proper care or follow label instructions, potentially resulting in injury to people or surrounding areas.

nematicide. A pesticide used to control nematodes.

nematode. Elongated, cylindrical, nonsegmented worms, commonly microscopic; some are parasites of plants or animals.

NIOSH. See *National Institute for Occupational Safety and Health*.

nitrile. An organic cyanide used to create synthetic rubber products. It is also used in pesticide products.

nonpoint source pollution. Pollution from pesticides or other materials that arises from their normal or accepted use over a large general area and extended period.

nonselective pesticide. A pesticide that has action against many species of pests rather than just a few; see also ***broad-spectrum pesticide.***

nontarget organism. Animals or plants within a pesticide-treated area that are not intended to be controlled by the pesticide application.

nymph. The larva of some insects such as mayflies, dragonflies, and grasshoppers that resembles the adult and develops into the adult insect directly, without passing through an intermediate pupa stage.

obsolete. No longer in use; outdated.

occasional pest. A pest that does not recur regularly, but causes damage from time to time as a result of changing environmental conditions or other factors.

offsite movement. Any movement of a pesticide from the location where it was applied, through drift, volatilization, leaching, runoff, crop harvest, blowing dust, or by being carried away on organisms or equipment.

oral. Through the mouth, one of the routes of entry of pesticides into the body.

oral notification. A method used to notify workers of pesticide applications on property where they are employed.

organic. A pesticide whose molecules contain carbon and hydrogen atoms; also, plants or animals that are grown without the use of synthetic fertilizers or pesticides.

organic matter. Any material that comes from living organisms; in soil, this would include decaying plant, microbial, and animal matter.

organism. Any living thing.

organophosphates. Organic molecules containing phosphorus commonly used as pesticides. Some are highly toxic to people; most break down in the environment very rapidly.

OSHA. Occupational Safety and Health Administration; the part of the U.S. Department of Labor that sets and enforces rules that keep people safe and healthy at work.

other ingredients. Ingredients other than the active ingredient in a pesticide formulation. Some may be toxic or hazardous to people.

output rate. The amount of pesticide mixture discharged by pesticide application equipment over a measured period of time.

overwinter. The process of passing through the winter season. Many living organisms survive harsh weather conditions as seeds, eggs, or in certain resting stages.

parasite. A plant or animal that derives all its nutrients from another organism; parasites often attach themselves to their host or invade the host's tissues; parasitism may result in injury or death of the host.

pathogen. A microorganism that causes a disease.

pellet. A pesticide formulation consisting of the dry active ingredient and other materials pressed into uniform-sized granules.

penetrate. To pass through a surface such as skin, protective clothing, plant cuticle, or insect cuticle; also, the ability of an applied spray to pass through dense foliage.

perennial. A plant that lives longer than 2 years; some may live indefinitely.

permeability. The ability of material (such as geological layers) to allow water and dissolved pesticides to move downward to groundwater freely.

persistent pesticide. A pesticide that remains active in the environment for long periods of time because it is not easily broken down by microorganisms or environmental factors.

personal protective equipment (PPE). Devices and garments that protect handlers from exposure to pesticides; these include coveralls, eye protection, gloves and boots, respirators, aprons, and hats.

pesticide. Any substance or mixture of substances intended for preventing, destroying, repelling, or mitigating any insects, rodents, nematodes, fungi, or weeds, or any other forms of life declared to be pests, and any other substance or mixture of substances intended for use as a plant regulator, defoliant, or desiccant.

pesticide deposition. See ***deposition***.

pesticide formulation. The pesticide as it comes from its original container, consisting of the active ingredient blended with other ingredients.

pesticide handler. See ***handler***.

pesticide residue. See ***residue***.

pesticide resistance. Genetic qualities of a pest population that enable individuals to resist the effects of certain types of pesticides that are toxic to other members of that species.

Pesticide Safety Information Series (PSIS). A series of informational sheets developed and distributed by the California Department of Pesticide Regulation pertaining to handling pesticides, personal protective equipment, emergency first aid, medical supervision, etc.

pesticide use hazard. The potential for a pesticide to cause injury or damage during handling or application.

pesticide use record. A record of pesticide applications made to a specific location.

Pest Management Guidelines. A series of crop-related publications from the University of California that provide research-based information on managing pests through chemical and nonchemical means. These guidelines are available through county University of California Cooperative Extension offices and can be accessed at www.ipm.ucanr.edu.

pest resurgence. See ***resurgence***.

pH. A value used to express relative acidity or alkalinity. Lower numbers indicate increasing acidity; higher numbers indicate increasing alkalinity.

photosynthesis. Process by which plants convert sunlight into energy.

physical controls. Activities designed to kill a pest or make the environment unsuitable for survival; physical controls include mowing, steam sterilization of soil, and installing screens or other barriers.

phytotoxic. Damaging to plants.

plantback restriction. A restriction that limits the commodity that can be grown in an area for a designated period of time after a certain pesticide has been used.

point source pollution. Pollution from pesticides or other materials that arises from spilling or dumping them in one location.

pollinators. Organisms that transfer pollen and fertilize plants; usually refers to bees.

population. A group of individuals of the same species occupying a distinct space and possessing characteristics (such as special adaptations for the habitat) that are unique to the group.

postemergence. Describes a pesticide that is applied after plants have emerged from the soil.

posting. Placing signs around an area to inform workers and the public that the area has been treated with a pesticide.

potential. Having or showing the capacity to become or develop into something in the future.

pour-on. A ready-to-use formulation or diluted mixture of pesticide for control of external parasites on livestock. The liquid is usually poured along the back of the animal.

powder. A finely ground dust containing active ingredient and other ingredients. This powder is mixed with water before application as a liquid spray.

power take-off (PTO). A special shaft connected to the rear, front, or side of a tractor and certain other types of equipment that uses the engine of the tractor or other equipment to power external devices such as sprayers, mowers, hydraulic pumps, etc.

PPE. See ***personal protective equipment***.

ppm. Parts per million.

precautionary statements. A section on pesticide labels listing human and environmental hazards and personal protective equipment requirements, as well as the product's specific effects on people and animals.

precipitation. Process by which solid particles settle out of a solution, such as a formulated pesticide in a spray tank. Also can refer to rain.

predacide. Pesticide used for control of predaceous mammals such as coyotes.

predator. An animal that attacks, kills, and eats other animals (prey), consuming several to many prey individuals in its lifetime.

preemergence. Describes a pesticide that is applied to a site before plants emerge from the soil.

preharvest interval. A period of time set by law that must elapse after a pesticide has been applied to an edible crop before the crop can be harvested legally.

pressure. The amount of force applied by the application equipment pump on the liquid pesticide mixture to force it through the nozzles.

pressure gauge. An instrument on liquid pesticide application equipment that measures the pressure of the liquid being expelled.

prey. An organism that is attacked, killed, and eaten by a predator.

private applicator. Individuals who apply pesticides on agricultural property under their control and for their own benefit or needs.

protective clothing. Garments of personal protective equipment that cover the body, including arms and legs.

psi. Pounds per square inch.

PTO. See ***power take-off***.

pupa (pl., pupae). In insects having complete metamorphosis, the resting life stage between larval and adult forms.

pyrethroid. A synthetic pesticide that mimics pyrethrin, a botanical pesticide derived from certain species of chrysanthemum flowers.

qualified trainer. A person who is a certified private or commercial applicator, agricultural pest control adviser, registered forester, agricultural biologist, or UC farm advisor, or who has completed a DPR-approved train-the-trainer course.

rate. The quantity or volume of liquid spray, dust, or granules that is applied to an area over a specified period of time.

rate controller. An electronic device installed on a pesticide sprayer that adjusts the spray volume automatically by adjusting spray pressure.

recombination. An occurrence in which a pesticide breaks down and combines with other chemicals in the environment to produce a different compound than what was originally applied.

recommendation. A written document prepared by a licensed pest control adviser that prescribes the use of a specific pesticide or other pest control method.

refuge. A place of safety. Also, an area near an application site that is left untreated to slow the development of pesticide resistance.

registrant. Company that obtained the registration of the pesticide product; also referred to as the manufacturer of the formulated product.

registration and establishment numbers. Identification numbers assigned by the U.S. EPA and the California Department of Pesticide Regulation found on pesticide labels.

regulations. Guidelines or working rules that a regulatory agency uses to carry out and enforce laws.

REI. See ***restricted-entry interval***.

rescue breathing. Also known as artificial respiration. It is given mouth-to-mouth to assist or restore breathing to a person overcome by pesticides.

reservoir. A population of pests within a local area; also, an organism harboring plant or animal pathogens.

residual action. The activity of a pesticide after it has been applied. Most pesticide compounds remain active several hours to several weeks or even months after being applied. Also known as residual activity.

residue. Traces of pesticide that remain on treated surfaces after a period of time.

resistance. See ***pesticide resistance*** or ***host resistance***.

respiration. Metabolic process in plants and animals in which, among other things, oxygen is exchanged for carbon dioxide or carbon dioxide is exchanged for oxygen.

Respirator Program Administrator. A person trained or experienced with respiratory equipment who has been selected to oversee a respiratory protection program.

respiratory equipment. A device that filters pesticide dusts, mists, and vapors to protect the wearer from respiratory exposure during mixing and loading, application, or while entering treated areas before the restricted-entry interval expires.

restricted-entry interval (REI). Period of time that must elapse between the application of a pesticide and when it is safe to allow people into the treated area without requiring that they wear personal protective equipment and receive early-entry worker training.

restricted materials permit. Permit, issued by county agricultural commissioners, that enables certified applicators to possess and apply restricted-use pesticides.

restricted-use pesticide. Highly hazardous pesticide that can be possessed or used only by commercial applicators who have a valid qualified pesticide applicator license or certificate or private applicators who have passed a written exam administered by the local agricultural commissioner.

restrictive statement. A statement on a pesticide label that restricts the use of that pesticide to specific areas or designated individuals.

resurgence. The sudden increase of a pest population after some event, such as a pesticide application.

reversible injury. A pesticide-related injury or illness that can be reversed through medical intervention and/or the body's healing process.

rhizome. An underground stem of certain types of plants.

rinsate. Liquid derived from rinsing pesticide containers or spray equipment.

rodenticide. A pesticide used to control rats, mice, gophers, squirrels, and other rodents.

rope wick applicator. A device used to apply contact herbicides onto target weed foliage with a saturated rope or cloth pad.

route of exposure. Any one of the four ways a pesticide gets onto or into the body: dermal (on or through the skin), ocular (on or in the eyes), respiratory (into the lungs), and oral (through the mouth); also known as route of entry.

rpm. Revolutions per minute.

runoff. Liquid spray material that drips from the foliage of treated plants or from other treated surfaces; also, rainwater or irrigation water that leaves an area and may contain trace amounts of pesticides.

Safety Data Sheet (SDS). An information sheet provided by a pesticide manufacturer describing chemical qualities, hazards, safety precautions, and emergency procedures to be followed in case of a spill, fire, or other emergency; formerly known as Material Safety Data Sheet (MSDS).

sampling. Collecting several examples of an organism from an area to determine pest identity. Techniques can also be used to get a sense of the size of the population in an area.

SCBA. See ***supplied air respirator***.

scouting. Collecting monitoring information in the field; used to detect and assess pest populations in an area.

secondary pest. An organism that becomes a serious pest only after a natural enemy, competitor, or primary pest has been eliminated through pest control.

selective pesticide. A pesticide that is effective against only a single or a small number of pest species.

sensor. A mechanical device that is sensitive to light, radiation, level, movement, heat, or other stimuli in the environment, and that provides a corresponding output.

service container. Any container designed to hold concentrated or diluted pesticide mixtures, including the sprayer tank but not the original pesticide container.

shingling. Clumping or sticking together of plant foliage caused by the force of a liquid spray; prevents spray droplets from reaching all surfaces of the foliage and may result in poor pest control.

sight gauge. A device on a pesticide sprayer or a configuration of the spray tank that permits the operator to view the level of liquid in the tank.

signal word. One of three words (DANGER, WARNING, CAUTION) found on every pesticide label to indicate the relative hazard of the chemical.

signs. The physical evidence of a pest's presence that can be seen on a host. For example, in plant disease, signs can include visible spores or fruiting bodies.

site. The area where pesticides are applied for control of a pest.

skin absorption. The passage of pesticides through the skin into the blood stream or other organs of the body.

skull and crossbones. The symbol on pesticide labels that indicates that the material is highly poisonous or poses specific, serious health or environmental hazards. Always accompanied by the signal word DANGER and the word POISON; see also ***DANGER-POISON***.

slurry. A watery mixture containing pesticide powder that leaves a thick coating of pesticide residue on treated surfaces.

soluble. Able to dissolve completely in a liquid.

soluble powder. A pesticide formulation in which the active ingredient and all other ingredients completely dissolve in water to form a true solution.

solution. A liquid that contains a dissolved substance, such as a soluble pesticide.

solvent. A liquid capable of dissolving certain chemicals.

species. A subdivision of a genus considered as a basic biological classification and containing individuals that resemble one another and may interbreed.

specific. Clearly defined or identified.

speed of travel. The speed that the operator moves the pesticide application equipment through the area being treated.

spore. A reproductive structure produced by some plants and microorganisms that is resistant to environmental influences.

spot treatment. A method of applying pesticides only in small, localized areas where pests congregate, rather than treating a larger, general area.

spray check device. A piece of equipment that measures and visualizes the output from the nozzles on a spray boom, providing rapid visualization of differences in output between nozzles.

spreader. An adjuvant that lowers the surface tension of treated surfaces to enable the pesticide to be absorbed.

statement of use classification. A special statement found on labels of some highly hazardous pesticides indicating their use is restricted to people who have been qualified through a certification process.

steam sterilization. A pest control method that kills bacteria and other living things in soil using steam.

sticker. An adjuvant used to prevent pesticides from being washed or rubbed off treated surfaces.

stunt. To restrict growth; in plants, a common symptom of disease or nematode infestation.

summer annuals. Annuals that germinate in the spring, mature and set seed in late summer, and die in the fall.

supplemental label. Additional instructions and information that are not found on the pesticide label because the label is too small, but which are legally considered to be part of the pesticide labeling.

supplied air respirator. A tightly fitting face mask that is connected by hose to an air supply such as a tank worn on the back of the person using the respirator or to an external air supply.

suppression. Pest management strategy that attempts to reduce pest numbers below an economic injury threshold or to a tolerable level.

surface coverage. The degree to which a spray or dust covers the surface of leaves or other objects being treated.

surface water. Water contained in lakes or ponds or flowing in streams, rivers, and canals.

surfactant. Surface-active agent; an adjuvant used to improve the ability of the pesticide to stick to and be absorbed by the target surface.

susceptible life stage. The life stage of a pest organism that is most likely to be affected by a pesticide used to control it; in general, insects are most susceptible during the larval or juvenile stage; weeds are usually most susceptible during the seedling stage.

suspension. Fine particles of solid material distributed evenly throughout a liquid such as water or oil.

sustain. To continue over time without losing effectiveness.

swath. The area covered by one pass of the pesticide application equipment.

swath width. The width of the area covered by spray droplets or granules as the application equipment moves through; the swath width must be measured to calibrate application equipment.

symptoms. Changes in the appearance of an organism due to the activities of a pest; for example, in plant disease, the appearance of lesions, cankers, or discolored leaves; also, any abnormal condition in people caused by a pesticide exposure that can be felt and described or can be detected by examination or laboratory tests.

synthetic. A product made artificially by chemical synthesis such as a pesticide or fabric.

system controller. See *rate controller*.

systemic pesticide. A pesticide that is taken up into the tissues of the organism and transported to other locations, where it will affect pests.

system monitor. A device that measures a sprayer's operating conditions such as travel speed, pressure, and/or flow rate and can send an alert when unexpected changes in application rates occur. Used in conjunction with GPS units, and spray, rate, or system controllers.

tailwater. The water that collects at the lower end of a field during or after irrigation.

tank mix. A mixture of pesticides or fertilizers and pesticides applied at the same time.

target. Either the pest that is being controlled or surfaces within an area that the pest will contact.

temperature inversion. See ***inversion***.

thickener. An adjuvant that increases the viscosity of the spray solution so that larger droplets are formed by the nozzles; used to control drift.

threatened species. An organism that is likely to become endangered in the foreseeable future.

threshold. The point at which the value of the damage caused by a pest exceeds the cost of controlling the pest, therefore making it practical to use a control method. Also known as economic injury or treatment threshold.

tolerance. The ability to endure the impact of a pesticide or pest without exhibiting adverse effects; also, the maximum amount of pesticide residue that is permitted on produce or other edible animal or agricultural crop products.

toxicity. The potential of a pesticide to poison an exposed organism.

toxicity category. The four classifications of pesticides that indicate the approximate level of hazard, indicated by the signal words DANGER-POISON, DANGER, WARNING, and CAUTION.

toxicity testing. A process in which known doses of a pesticide are given to groups of test animals and the results are observed.

toxicology. The study of toxic substances on living organisms.

trade name. See ***brand name***.

training record. Document signed by the trainer, employer, and trainee that records the dates and types of pesticide safety training received.

translocate. The movement of pesticides from one location to another within the tissues of a plant.

transmit. To cause something (like a disease) to pass from one place or organism to another.

transport. To carry somebody or something from one place to another, usually in a vehicle.

treated surface. The surface of plants, soil, or other items that were contacted with pesticide spray, dust, or granules for the purpose of controlling pests.

treatment area. See ***site***.

triazines. A large family of chemicals used to control both broadleaf and grassy weeds, either before or after they emerge, by inhibiting photosynthesis.

triple-rinse. Performing three times the process of partially filling an empty pesticide container, replacing the lid, shaking the container, then emptying its contents into the spray tank.

tuber. An enlarged, fleshy, underground stem.

UC ANR. University of California Agriculture and Natural Resources.

UC IPM. University of California Statewide Integrated Pest Management Program.

ultra-low-volume (ULV). A pesticide application technique in which very small amounts of liquid spray are applied over a unit of area; usually ½ gallon or less of spray per acre in row crops to about 5 gallons of spray per acre in orchards and vineyards.

ULV. See ***ultra-low-volume***.

uniform. Always the same in quality, character, degree, or manner, as in uniform pesticide distribution, uniform size, or uniform mixture.

unloader. A sensitive, valve-like mechanism used on high-pressure applicators that diverts the liquid back into the tank when nozzles are shut off to prevent a rapid buildup of pressure in the system that would possibly damage the pump. When the flow to the nozzles is turned back on, the unloader quickly restores pressure to nozzles.

unregistered crop. Any crop that is not listed on the pesticide label; pesticides can be applied only to crops that are specifically listed on the label.

unregistered site. Any site, such as a right-of-way or pond, that is not listed on the pesticide label; pesticides may be applied only to listed or registered crops or sites.

U.S. Environmental Protection Agency (U.S. EPA). The federal agency responsible for regulating pesticide use in the United States.

use restrictions. Special restrictions included on the pesticide label or incorporated into state or local regulations that specify how, when, or where a specific pesticide may be used.

vapor pressure. The pressure exerted by a material in its gaseous form.

variables. Factors that differ from place to place, or situation to situation. Also, the part of a mathematical equation that does not have a fixed numerical value.

variety. In plants, naturally occurring variants within a subspecies, or strains produced through breeding programs. Also, a collection of varied things, often belonging to the same group.

vector. An organism, such as an insect, that can transmit a pathogen to plants or animals.

vegetative. Relating to or typical of vegetation, plants, or plant growth. Also, asexual reproduction.

vertebrates. The group of animals that have an internal skeleton and segmented spine, such as fish, birds, reptiles, and mammals.

viroid. A microorganism that is much smaller than a virus and is not enclosed in a protein coat; some viroids produce disease symptoms in certain plants.

virus. A very small organism that multiplies in living cells and is capable of producing disease symptoms in some plants and animals.

viscosity. A physical property of a fluid that affects its flowability; more-viscous fluids flow less easily and produce larger spray droplets.

VOC. See ***volatile organic compound***.

volatile. Able to pass from liquid or solid into a gaseous stage readily at low temperatures.

volatile organic compound. An organic compound that evaporates at a relatively low temperature and contributes to air pollution.

volatilization. The process of a liquid or solid passing into a gaseous stage.

WARNING. The signal word used on labels of pesticides considered to be moderately toxic or hazardous based on their toxicity; they usually have an oral LD_{50} between 50 and 500 and a dermal LD_{50} between 200 and 2,000.

watershed. An area of land that drains its surface water into a defined watercourse or body of water.

water-soluble concentrate. A liquid pesticide formulation that dissolves in water to form a true solution.

weather. The state of the atmosphere (temperature, humidity, precipitation, wind conditions) over a short period of time (a day or week) at a specific site.

weather station (electronic). A device that consists of an electronic data logger, sensors, a power supply, an environmental enclosure, and a support structure.

weed. A plant that interferes with human activities, results in economic loss, or is otherwise undesirable.

wettable powder. A pesticide formulation consisting of an active ingredient that will not dissolve in water, combined with mineral clay and other ingredients and ground into a fine powder.

winter annuals. Annuals that germinate from late fall to early winter, mature and set seed in late winter or early spring, and die in early summer.

work clothing. Garments such as long-sleeved shirts, short-sleeved shirts, long pants, short pants, shoes, and socks; work clothing is not considered personal protective equipment, although pesticide product labeling or regulations may require specific work clothing during some activities.

Worker Protection Standard (WPS). The 1992 amendment to the Federal Insecticide, Fungicide, and Rodenticide Act (FIFRA) that makes significant changes to pesticide labeling and mandates specific training of pesticide handlers and workers in production agriculture, commercial greenhouses and nurseries, and forests; updated in 2017.

REFERENCES

Akesson, N. B., and W. E. Yates. 1964. Problems related to application of agricultural chemicals and resulting drift residues. *Annual Review of Entomology* 9:285–315, table 1.

Bauer, E., ed. 2005. *Agricultural Pest Control: Plant*. Lincoln: University of Nebraska.

Bird, G. W. 2003. Role of integrated pest management and sustainable development: Historical development of pest management programs. In K. M. Maredia, D. Dakouo, and D. Mota-Sanchez, eds., *Integrated Pest Management in the Global Arena*. Wallingford, UK: CABI Publishing.

Blecker, L. A., and J. M. Thomas. 2012. *National Soil Fumigation Manual*. Fairfax, VA: National Association of State Departments of Agriculture Research Foundation (NASDARF).

California Department of Pesticide Regulation. 2014. *A Community Guide to Recognizing & Reporting Pesticide Problems*. DPR website, cdpr.ca.gov/docs/dept/comguide/commty_guide.pdf.

——. 2015. *Using Pesticides in California*. DPR website, cdpr.ca.gov/docs/dept/comguide/using_excerpt.pdf.

DiTomaso, J. M., and E. A. Healy. 2007. *Weeds of California and Other Western States*. 2 vols. Oakland: University of California Division of Agriculture and Natural Resources Publication 3488.

Dreistadt, S. H. 2016. *Pests of Landscape Trees and Shrubs: An Integrated Pest Management Guide*. 3rd ed. Oakland: University of California Division of Agriculture and Natural Resources Publication 3359.

Dubrovsky, N. M., C. R. Kratzer, L. R. Brown, J. M. Gronberg, and K. R. Burow. 1998. *Water Quality in the San Joaquin-Tulare Basins, California*, 1992-95. U.S. Geological Survey Circular 11595.

Feldmann, R. J., and H. I. Maibach. 1967. Regional variation in percutaneous penetration of 14C cortisol in man. Journal of Investigative Dermatology 48(2) (Feb):181–183.

Flint, M. L. 1998. *Pests of the Garden and Small Farm*. 2nd ed. Oakland: University of California Division of Agriculture and Natural Resources Publication 3332.

——. 2012. *IPM in Practice*. 2nd ed. Oakland: University of California Division of Agriculture and Natural Resources Publication 3418.

FRAC (Fungicide Resistance Action Committee). Website, frac.info.

Hickman, G. W. 2004. *Pest Notes: Lizards*. Oakland: University of California Division of Agriculture and Natural Resources Publication 74120. UC ANR website, ipm.ucanr.edu/PMG/PESTNOTES/pn74120.html.

Hooven, L., R. Sagili, and E. Johansen. 2013. *How to Reduce Bee Poisoning from Pesticides*. Corvallis: Oregon State University PNW 591. OSU Extension website, https://catalog.extension.oregonstate.edu/files/project/pdf/pnw591.pdf.

HRAC (Herbicide Resistance Action Committee). Website, hracglobal.com.

IRAC (Insecticide Resistance Action Committee). Website, irac-online.org.

Kansas Department of Agriculture. 1990. *Chemigation in Kansas*. Topeka: Kansas Department of Agriculture.

Klingman, G. C., and F. M. Ashton. 1975. *Weed Science Principles and Practices*. New York: Wiley.

Kranz, W., C. Burr, J. Hay, J. Schild, and D. Yonts. 2008. *Using Chemigation Safely and Effectively: Training Manual*. Historical Materials from University of Nebraska-Lincoln Extension Paper 915.

Lovatt, C. n.d. Plant growth regulator strategies and avocado phenology and physiology. California Avocado Growers website, californiaavocadogrowers.com/sites/default/files/documents/PRG-Strategies-and-avocado-phenology.pdf.

McDonald, S. A. 1991. *Applying Pesticides Correctly: A Guide for Private and Commercial Applicators*. Washington, DC: US EPA and USDA.

McKenry, M. V., and P. A. Roberts. 1985. *Phytonematology Study Guide*. Oakland: University of California Division of Agriculture and Natural Resources Publication 4045.

National Institute for Occupational Safety and Health. 1985. *Occupational Safety and Health Guidance Manual for Hazardous Waste Site Activities*. Washington, DC: NIOSH.

——. 1998. *Setting Up Emergency Decontamination Facilities*. Washington, DC: NIOSH.

Occupational Safety and Health Administration. 2012. *Hazard Communication Standard: Safety Data Sheets*. OSHA Brief. OSHA website, osha.gov/Publications/OSHA3514.pdf.

——. 2014. *Occupational Heat Exposure*. OSHA website, osha.gov/SLTC/heatstress.

Perry, E. J., and A. T. Ploeg. 2010. *Pest Notes: Nematodes*. Oakland: University of California Division of Agriculture and Natural Resources Publication 7489. ipm.ucanr.edu/PMG/PESTNOTES/pn7489.html.

Pfeiffer, M. 2010. Ground Water Ubiquity Score (GUS). Tucson, AZ: Pesticide Training Resources. PTR website, ptrpest.com/pdf/groundwater_ubiquity.pdf.

Platt, H. D. 1953. Pictorial key to some common adult cockroaches. Atlanta: U.S. Department of Health, Education, and Welfare, Public Health Service Communicable Disease Center.

Randall, C., et al., eds. 2008. *National Pesticide Applicator Certification Core Manual*. Washington D.C.: National Association of State Departments of Agriculture Research Foundation.

Salmon, T. P., and R. A. Baldwin. 2009. *Pest Notes: Pocket Gophers*. Oakland: University of California Division of Agriculture and Natural Resources Publication 7433. UC IPM website, ipm.ucanr.edu/PMG/PESTNOTES/pn7433.html.

Salmon, T. P., and W. P. Gorenzel. 2010a. *Pest Notes: Ground Squirrels*. Oakland: University of California Division of Agriculture and Natural Resources Publication 7438. UC IPM website, ipm.ucanr.edu/PMG/PESTNOTES/pn7438.html.

——. 2010b. *Pest Notes: Rabbits*. Oakland: University of California Division of Agriculture and Natural Resources Publication 7447. UC IPM website, ipm.ucanr.edu/PMG/PESTNOTES/pn7447.html.

Salmon, T. P., D. A. Whisson, and R. E. Marsh. 2006. *Wildlife Pest Control around Gardens and Homes*. 2nd ed. Oakland: University of California Division of Agriculture and Natural Resources Publication 21385.

Saw, L., J. Shumway, and P. Ruckart. 2011. Surveillance data on pesticide and agricultural chemical releases and associated public health consequences in selected US states, 2003–2007. *Journal of Medical Toxicology* 7:164–171.

Schwankl, L. 2015. Microirrigation systems. UC Davis Fruit and Nut Research and Information website, fruitsandnuts.ucdavis.edu/files/73686.pdf.

Schwankl, L. J., and T. Prichard. 2001. Uniform chemigation in tree and vine microirrigation systems. Oakland: University of California Division of Agriculture and Natural Resources Publication 21599.

Stetson, D. I., and R. A. Baldwin. 2010. *Pest Notes: Birds on Tree Fruits and Vines*. Oakland: University of California Division of Agriculture and Natural Resources Publication 74152. UC IPM website, ipm.ucanr.edu/PMG/PESTNOTES/pn74152.html.

University of California Statewide Integrated Pest Management Program. Website, ipm.ucanr.edu.

U.S. Environmental Protection Agency. 1994. *Pest Smart Update*. Washington, D.C.: Publication EPA-733-N-94-001.

——. 2014. *Label Review Manual, Chapter 10: Worker Protection Label*. EPA website, epa.gov/sites/production/files/2014-07/documents/chapter10-final-fd-jr.pdf.

——. 2015. *Protecting Bees and Other Pollinators from Pesticides*. EPA website, epa.gov/pesticides/ecosystem/pollinator/bee-label-info-lrt.pdf.

Vertebrate Pest Control Handbook Online. Vertebrate Pest Control Research Advisory Committee website, vpcrac.org/about/vertebrate-pest-handbook.

INDEX

An *italic* "f" after a page number, such as 44*f*, refers to a photo or an illustration; an *italic* "t", as in 168*t*, refers to a table.

A

abiotic disorders, 23, 27
abrasive pesticides
 formulations, 35–37*t*, 156*t*
 nozzle selection, 114
 pump selection, 111–113, 156*t*
accidental release measures, Safety Data Sheets, 50
accidents. *See* emergency situations; environmental hazards; exposure to pesticides
active ingredients (a.i.), 35, 43, 44*f*, 146–147, 168*t*
acute toxicity, 72–73
adjuvants, 30, 38, 159, 165
adsorption process, 56, 58–59
agitators, 113
agricultural use requirements, pesticide label, 45*f*, 47
air blast sprayers, 119*t*, 136
air currents, drift potential, 57, 59–60, 105*t*
air gaps, filling procedure, 61, 104, 133, 166
air-purifying respirators, 86*t*, 87*f*
air-supplying respirators, 86*t*, 177
algae, 13–14, 32*t*
aluminum and monel nozzles, 114
amaranth (*Amaranthus retroflexus*), 15*f*
Anagyrus pseudococci (parasitic wasp), 3
annual bluegrass, 15
annual weeds, 12, 13*f*
ants, 3
aphids, 19
application equipment
 cleaning cautions, 58, 59, 70, 106
 components, 110–111
 inspection guidelines, 102, 111–114, 117, 122–123
 maintenance guidelines, 117, 121–124, 142
 types compared, 118–121*t*
 See also nozzles; pumps
application exclusion zones (AEZs), 96
application methods
 effectiveness of, 158–159
 exposure prevention, 68, 69
 follow-up monitoring, 169
 label requirements, 45*f*, 46–47, 63*f*
 nontarget species protection, 167–169
 residue control, 60–61
 safety planning, 96–100
 timing considerations, 60, 62, 158, 166
 types compared, 118–121*t*
 See also calibration methods; timing considerations; weather conditions
application rates, 72, 141, 146, 149
aprons, 82
aquifers, 57–59, 61, 106
 See also groundwater
armyworms, 20
arthropod pests, 16, 154, 161
atmosphere-supplying respirators, 86*t*
Atrazine, 32, 33

B

backflow prevention, 59, 61, 104, 166
backpack sprayers, 62, 118*t*, 119*t*, 130, 131–132, 137–138
bacteria, 24–26, 32*t*
bacterial spot, 24*f*
bacterial wilts, 25
bagrada bugs, 18–19
bait treatments, 38*f*, 121*t*
band applications, 118*t*, 136
barnyard grass, 11*t*, 12, 15
basal applications, 119*t*
bees. *See* honey bees
beetles, 4*f*, 16*t*, 26, 27
beneficial species, 17, 18, 19, 21, 24
 See also natural enemies; nontarget species
bermudagrass, 14*f*
biennial weeds, 12
bioaccumulation of pesticides, 62, 63*f*
biological control, 2–3, 4
 See also natural enemies
biology of pests, 10, 154, 158–159
birds, 12, 21, 32*t*, 168–169
body parts, absorption rates, 70–71
 See also exposure to pesticides
boom-type sprayers, 118–120*t*, 134–135, 136, 137*f*
boots, protective, 84, 88
brand names, pesticides
 groupings of, 32–35
 label information, 43, 44*f*
broadcast applications, 119*t*
broadleaves, 12, 15–16
brown rot of stone fruits, 23*f*
buffer zones
 drift reduction, 165
 label information, 106
 runoff reduction, 166, 168
 sensitive area protection, 105, 106, 166
 warning signs, 98
butterflies, 20

C

cabbage looper, 11*f*
calibration methods
 accuracy importance, 104–105, 128, 141
 adjusting after maintenance, 124
 calculating pesticide amounts, 138–139, 146
 changing output rates, 141–142
 controller/monitor systems, 149–150
 dilution calculations, 146–149, 150*f*
 flow rate measurement, 130–134, 135*f*, 142–145
 residue control, 61

swath width measurements, 134–138, 145
tank capacity measurement, 129–130
tools for, 128–129
travel speed measurement, 130, 141, 142
worksheet, 140*f*
California Code of Regulations
decontamination facilities, 196–197
section references listed, 188
See also legal requirements
cankers, 24*f*, 25
carbamates, 33*t*
cartridge respirators, 86*f*, 89–90
Category I-IV pesticides, 31
CAUTION signal
pesticide labels, 31, 46
toxicity category, 31, 156
centrifugal pumps, 113
ceramic nozzles, 114
chemical controls, IPM programs, 5
chemical names
groupings of, 32–35
label requirements, 44*f*, 46
chemical properties information, Safety Data Sheets, 50
chemical-resistant clothing
cleaning/disposal procedures, 87, 90
types, 80–84
See also personal protective equipment (PPE)
chemigation, 98, 110, 121*t*, 123–124
chewing lice, 17
children, pesticide hazards, 68–69, 71
chinch bugs, 18–19
chlorothalonil, 147*f*
chronic toxicity, pesticide exposure, 73
cicadas, 19–20
Class I disposal facilities, 122, 180
Class II landfill, 103
classical biological control, 4
classification statements, pesticide labels, 43, 44*f*
cleaning procedures
application equipment, 59, 106, 117, 121–124, 159
environmental hazards, 58, 59, 166
protective equipment, 85, 87–90
training requirements, 78
vineyard case study, 3
See also disposal procedures
cleanup of spills, 180, 181–182
closed mixing systems, 91–92, 102
clothing, protective, 80–84, 85, 87–88, 90, 106
See also personal protective equipment (PPE)
Code of Regulations
decontamination facilities, 196–197
section references listed, 188
See also legal requirements
compatibility problems during mixing, 159–161
contact lenses, 84
contact pesticides, 34
containers, pesticide
disposal procedures, 60, 103
storage requirements, 68–69, 71, 72, 102
triple-rinsing, 103, 104*f*
contents list, pesticide labels, 44*f*, 46
controllers, 112, 115*t*, 116*t*, 118*t*, 149–150
Convolvulus arvensis (field bindweed), 15*f*
corrosion of equipment
inspection guidelines, 117, 123
nozzles, 114
preventing, 122
tanks, 111
cost factors, pesticides, 157
coveralls, 31, 81–82
crop restrictions, pesticide label requirements, 45*f*, 46–47
cultural controls, IPM programs, 5

D

DANGER-POISON signal, 31, 44*f*, 46
DANGER signal
pesticide labels, 31, 44*f*, 46
sign posting, 51*f*, 98, 99*f*, 102
toxicity category, 31, 156
decontamination facilities, 52, 102, 178–179, 196–197
depredation permits, 21
dermal exposure. *See* skin exposure
diaphragm pumps, 111, 112
Diazinon 50W, 35
dichotomous keys, 10–11
dicots (broadleaves), 12, 15–16
dilution calculations, 146–149, 150*f*
dip applications, 120*t*
dipsticks, 130, 133
direct-channel contamination, 59
directed-spray applications, 118*t*
directions for use, label requirements, 45*f*, 46–47
diseases, 23–27
disorders, 23, 27
disposal procedures
cleaning water, 122, 166
containers, 60, 103, 166
environmental hazards, 58, 59, 60, 167
glove liners, 83
leftover pesticides, 122, 159
pesticide label directions, 45*f*, 47
protective clothing, 83, 87
Safety Data Sheet instructions, 50
spill accidents, 180, 182
water supply protection, 166
drift events
application timing, 98
causes, 57, 59–60, 162–163
injury potential, 68
preventing and reducing, 142, 162–165, 166
sensitive area protection, 105–106
weather factors, 105*t*, 163*t*
drizzle applications, 118*t*
droplets, spray
calibration adjustment effects, 142
drift control, 162–163, 164–165
size charts, 115*f*, 164*t*
See also nozzles
dry flowables, 156*t*, 159
dust formulations
applicators, 123
characteristics, 37*t*, 156*t*
drift potential, 163*t*
dust/mist masks, 86*t*, 86*f*

E

early-entry workers, 47, 78, 96
ecological information, Safety Data Sheets, 50
emergency numbers, 174
emergency situations
decontamination facilities, 178–179, 196–197
fires, 182
first aid procedures, 174, 176–178
heat-related illness, 74, 82, 176
incident review process, 184
label information, 44*f*, 46
misapplication accidents, 183–184
recognizing symptoms, 176
response planning, 85, 96, 101, 174–175, 184
spill accidents, 179–182
stolen pesticides, 182
training requirements, 79
employer responsibilities, personal safety, 79
emulsifiable concentrates, 36*t*, 156*t*
enclosed cabs, 91
endangered species, 62
engineering controls, 50, 90–92
environmental hazards
cleaning cautions, 58, 59, 166
importance of calibration accuracy, 128
label information, 45*f*, 46
nontarget species, 57–58, 62–63
offsite movement factors, 56–60, 61
residue buildups, 60–61
Safety Data Sheet, 50
sensitive areas, 61–62, 105–106
storage cautions, 102
erosion, offsite movement, 56, 58, 61, 166
Erwinia rubrifaciens (bacteria), 24*f*, 25*f*
establishment number, pesticide labels, 44*f*, 46
ethoprop, 148*f*
explosion hazards, label requirement, 45*f*, 46
exposure to pesticides
effects and symptoms, 72–74
first aid procedures, 176–178
injury routes, 30, 70–72
limits of protective equipment, 90
pesticide label emergency information, 44*f*, 46
possibilities for, 68–69
Safety Data Sheet information, 50
See also livestock
eye exposure
first aid procedures, 177
process of, 70*f*, 71
eyewear, protective, 71, 84, 86, 89, 132

F

face shields, 84–85, 89
facial hair, respirator guidelines, 85, 87*f*
facilities maps, 175
fiberglass tanks, 111
field bindweed *(Convolvulus arvensis)*, 15
field incompatibility, causes, 159
field posting requirements, 51
fieldworkers
exposure potential, 69
protection planning, 51, 96, 98, 100
training requirements, 78–79, 190–191
filter screens, 110, 113, 117, 159
fire blight, 24*f*, 25*f*
firefighting measures, Safety Data Sheets, 49
fire hazards
emergency responses, 181, 182
pesticide label requirement, 45*f*, 46
storage facilities, 102
first aid procedures
eye exposure, 177
inhalation of pesticides, 177–178
pesticide label statement, 44*f*, 46
Safety Data Sheet information, 49
skin exposure, 176–177
swallowed pesticides, 178
fish, 32*t*, 168
fit testing respirators, 71, 85
flat-fan nozzles, 113, 116*t*
flies, 20–21
flooding nozzles, 113, 116*t*
flowables, 35*t*, 36*t*, 156*t*, 159
flowmeters, 129–130, 133
flow rates, measuring
granule applicators, 142–145
liquid applicators, 129, 130–134, 135*f*, 142–143
foliar treatments, 120*t*
follow-up monitoring, 3, 169
footwear, protective, 84, 88
formulations, pesticide
dilution calculations, 146–149, 150*f*
equipment wear, 114, 121
label information, 44*f*, 46
types, 35–38, 156*t*
fumigants, 62
fungi, 23–24, 32*t*
fungicides, 32–34

G

galls, 25
gas masks, 86*t*
gloves
during calibration process, 131–132, 142
cleaning procedures, 85, 88
during mixing process, 103
during nozzle cleaning, 121
types of, 82–83
glyphosate, 148*f*
gnats, 20–21
Goaltender, 35
goggles, 84–85, 89
gophers, 22
GPS units, 130, 158
granule applicators
calibrating, 142–146
maintenance, 123
granule formulations
bird perils, 168–169
characteristics, 35*t*, 37*t*, 156*t*
dilution calculations, 148*f*
grasses, 11*t*, 12, 14–15
greenhouses, pesticide hazards, 69
ground squirrels, 22–23
groundwater
contamination, 58–59, 61, 168
protection, 102, 106, 165–166
Groundwater Protection Area (GWPA) maps, 106

H

habitat modification, 6
half-life of pesticides, 30, 50
handlers
  California Code of Regulations, 188, 196–197
  defined, vi
  See also training requirements
handling precautions, Safety Data Sheets, 50
hand-operated sprayers, 110, 118–121*t*, 130–132, 145–146
hand washing, 71, 72, 78, 106
harvest intervals, 45*f*, 47, 68
hazards of pesticides
  label information, 44–45*f*, 46
  Safety Data Sheet information, 49
  toxicity factors/categories, 30–31
  See also environmental hazards; exposure to pesticides
heat-related illness, 74, 82, 176
herbicides, 32–34
high-pressure sprayers, 119*t*, 132–133
honey bees
  hazard potential, 62, 168*t*
  label information, 63*f*
  notification requirements, 62
  protection methods, 167–168
  Safety Data Sheet, 50
hose-end sprayers, 118*t*
host resistance, 6
human hazards. *See* exposure to pesticides

I

identification requirement, Safety Data Sheets, 48
identifying pests
  accuracy importance, 2, 10
  diseases and disorders, 23–27
  invertebrates, 16–21
  resources for, 10–12
  vertebrates, 21–23
  weeds, 12–16
  See also selection factors, pesticides
impregnate formulations, 38*f*
incompatibility of pesticides, 50, 159, 160
ingredients information, Safety Data Sheets, 49
inhalation hazards
  first aid, 177–178
  formulation comparisons, 156*t*
insecticides, 32–34
insects, 16–21, 32*t*
  See also selection factors, pesticides
inspection procedures
  application equipment, 102, 111–114, 117, 122–123
  respiratory equipment, 89
  transport vehicles, 100
integrated pest management (IPM), 1–7, 62
inversions, temperature, 105*t*, 106*f*, 163*t*, 165
invertebrate pests, 16–21
invert emulsion formulations, 36*t*
IPM (integrated pest management), 1–7, 62
irrigation systems, pesticide use, 61, 110, 123–124

K

key code card, glove, 83*f*
keys, identification, 11–12
Knock'em down 3SL, 43–47
knowledge expectations, for certification, vii

L

labels, pesticide
  availability requirements, 42, 63, 96
  format requirements, 42
  reading recommendations, 43
  storage visibility, 102
labels, pesticide information
  active ingredients, 35, 43, 46, 147
  buffer zones, 168
  classification statement, 43
  cleaning instructions, 122
  directions for use, 42, 45, 46–47, 122
  formulation type, 47
  hazards, 47
  mixing system requirements, 91
  mode of action group, 34*f*
  pollinator protection, 163*t*
  personal protective equipment requirements, 44, 78, 83*f*
  sample with explanations, 43–47
  sensitive area protection, 167
  signal word, 46
  soil texture, 59
  specialized equipment requirements, 111*f*
  toxicity categories, 31
  toxicity warnings, 73
  water contamination warnings, 165–166
LC_{50} (lethal concentration) of pesticides, 30
LD_{50} (lethal dose) of pesticides, 30
  leaching
  hazards, 68, 105, 106, 120*t*
  process, 57, 58–59
  See also groundwater
leaffooted bugs, 18–19
leafhoppers, 19–20
leafminers, 20–21
leaks, spills of pesticide, 179–182
legal requirements
  calibration accuracy, 128
  employee notifications, 51–52
  labels, 42–47
  permits, 51
  personal safety, 78–79
  protective equipment, 71, 78, 82–83, 84
  public notifications, 61–62
  registration, 42
  Safety Data Sheets, 48–50
  safety training, 78–79
lethal concentration (LC_{50}) of pesticides, 30
lethal dose (LD_{50}) of pesticides, 30
lice, 17–18
life cycle of pests, 10, 154, 158–159
liquid formulations
  dilution calculations, 146–147, 148*f*, 150*f*
  types of, 35–36*t*
livestock
  exposure possibilities, 60, 62, 69, 169
  label requirements, 45*f*, 46–47
  symptoms of exposure, 62–63

loading procedures. *See* mixing and loading
low-pressure sprayers, 118*t*, 131–132, 133*f*
lygus bugs, 18–19

M

maggots, 20
maintenance guidelines
  application equipment, 111–114, 121–124, 142
  transport vehicles, 100
mammal pests, 12, 22–23, 32*t*
  See also selection factors, pesticides
manufacturer information, 44*f*, 46
maps
  emergency planning, 175
  permit applications, 51
mealybugs, 2–3, 19
mechanical agitators, 113
mechanical controls, IPM programs, 5
medical care, 174–175, 176–178, 188
metal tanks, 111
microencapsulated formulations, 37*t*
microorganisms, 10, 63*f*, 157
midges, 20–21
misapplication of pesticides
  environmental hazards, 60, 63
  types, 72, 183–184
misuse statement, pesticide labels, 45*f*, 47
mites, 17, 32*t*
  See also selection factors, pesticides
mixing and loading
  closed systems, 91–92, 102
  compatibility problems, 159–160
  dilution calculations, 146–149, 150*f*
  exposure possibilities, 68, 69
  eyewear protection, 167
  general guidelines, 159
  point source pollution potential, 58
  preliminary testing, 96
  safety guidelines, 102–104
  sensitive area protection, 167
  site selection, 61
  water supply protection, 166
modes of action, 33–34, 158–159
molluscicides, 32–33
Monilinia spores, 23*f*
monitoring pests, 3, 154, 161–162, 169
monitors and controllers, 149–150
monocots (grasses), 11*t*, 12, 14–15
mosaic viruses, 25*f*
mosquitoes, 20–21
moths, 20
movement of pesticides. *See* offsite movement of pesticides

N

natural enemies
  case studies, 2–3, 5
  IPM programs, 2–3, 4
  pesticide hazards, 62, 63
  protection methods, 5, 157
nematodes, 10, 32*t*
  See also selection factors, pesticides
NIOSH-approved respirators, 44, 86*f*
nonpoint source pollution, 58
nonselective pesticides, 62, 156
nontarget species
  pesticide hazards, 57–58, 61, 62–63
  pesticide label requirement, 45*f*, 46
  protection methods, 157, 158, 167–169
  toxicity testing, 30
notification requirements
  pre-application, 98–100
  public communication, 96, 98
  restricted materials use, 51
  school sites, 61–62
  warning signs, 69*f*, 98, 99, 102
nozzles
  drift control, 164
  inspection procedures, 113–114, 130–131, 158
  maintenance guidelines, 121, 122*f*
  measuring output, 130–131, 132, 133–134*f*
  monitors and controllers, 149–150
  output adjustments, 142
  selection guidelines, 114–116
  types, 115–116

O

offsite movement of pesticides, 56–60, 61, 162–165
oil-soluble pesticides, 69, 70, 87, 111–112
oral exposure
  first aid procedures, 178
  process of, 70*f*, 71–72
  risks for children, 68–69, 71
organic matter, 59, 157
organophosphates, 33*t*
oscillating boom sprayers, 119*t*
other ingredients, in pesticides, 35, 43, 44*f*, 46, 147*f*
output rates, granule applicators, 142–145
output rates, liquid applicators
  change adjustments, 142–143
  measuring, 129, 130–134, 135*f*
oxytetracycline, 150*f*

P

packaging of pesticides, 37*t*, 92, 102–103
parathion, 70–71
particle/dust drift, 162, 163*t*
  See also drift events
particulate respirators, 86*t*
parts per million dilutions, 149
patch spray method, 118*t*
pathogens, 4, 10, 12, 23–27
peach leaf curl, 23*f*
pellet formulations, 37*t*
perennial weeds, 12, 13*f*
permits
  depredation, 21
  restricted materials, 51
persistence of pesticides
  offsite movement potential, 56, 57–59, 60
  worker exposure potential, 69
personal protective equipment (PPE)
  aprons, 82
  boots, 84, 88
  during calibration process, 128–129, 131–132, 142
  cleaning procedures, 85, 87–90, 106

clothing, 80–84, 85, 87–88, 90, 106
coveralls, 31, 81–82
emergency situations, 174, 177, 179–180, 181
employer responsibilities, 79, 82–83, 90
engineering controls, 90–92
eyewear, 84–85, 103
face shields, 84–85, 89
footwear, 84, 88
gloves, 82–83
goggles, 84–85, 89
heat-related illness issues, 74, 82
legal requirements, 71, 78
limits of effectiveness, 90
for mechanics/cleaners, 69
during mixing and loading, 103, 160
pesticide label requirements, 44–45*f*, 46, 47
purpose, 68, 71, 78
respiratory equipment, 85, 86–87*f*, 91
Safety Data Sheet information, 50
safety training, 78–79
suits, 82
pesticide resistance, 72, 128, 159, 161–162
pesticides, overview
definitions, 30, 187
groupings of, 32–38
toxicity factors, 30–31
See also specific topics, e.g., environmental hazards; mixing and loading; selection factors, pesticides
Pesticide Safety Information Series (PSIS), vi, 79
A-3, 92
A-5, 71
A-8 and A-9, 52
pesticide safety training, 78–79, 101, 190–195
pest identification. *See* identifying pests
pest resurgence, 62, 158
pH levels
pesticide breakdown, 157
testing and adjusting, 160, 161*f*
phone numbers, emergency situations, 174
phylloxerans, 19–20
physical controls, IPM programs, 5
physical properties information, Safety Data Sheets, 50
phytotoxic injury, 167
pictograms, in Safety Data Sheet, 49
piston pumps, 112
planning guidelines
emergency situations, 85, 96, 101, 174–175, 184
pesticide applications, 96–100
plantback restrictions, 47
plastic nozzles, 114
pocket gophers, 22
point source pollution, 58
poisoning symptoms, 68, 70, 73–74, 176–178
pollinators, 62, 63*f*
See also honey bees
posting pesticide warnings, 51–52, 98–100, 188
powdery mildew, 24*f*
power dusters, 119*t*, 120*t*
PPE. *See* personal protective equipment (PPE)
precautionary statements
pesticide labels, 44–45*f*, 46, 73, 91*f*, 102
Safety Data Sheets, 49
predators, 62, 63*f*, 169
preharvest intervals
label requirements, 45*f*, 47
pesticide selection, 157
pressure gauges, 110, 129, 131, 132
pressure regulators, 110, 132, 142
protective equipment. *See* personal protective equipment (PPE)
psyllids, 19–20
pumps, 110, 111–112
pyrethroids, 33*t*

R

rate controllers, 112, 115*t*, 116*t*, 118*t*, 149–150
reactivity information, Safety Data Sheets, 50
recordkeeping
application planning checklist, 97
flow rate measurements, 132
pesticide use, 52
respiratory protection program, 85
tank capacity markings, 130
training forms, 79, 190–195
recycling pesticide containers, 103
redroot pigweed, 15
registration of pesticides, 42, 46, 47*f*
regulatory information, Safety Data Sheets, 50
See also legal requirements
reporting requirements, 52, 180–181, 182
residues, pesticide
application equipment, 106, 122, 129
environmental buildups, 60–61, 169
resistance, pest, 72, 128, 159, 161–162
respiratory equipment
air-purifying respirators, 86*t*, 87*f*
air-supplying respirators, 86*t*, 177
atmosphere-supplying respirators, 86*t*
canister respirators/gas masks, 86*t*
cartridge respirators, 86*f*, 89–90
cleaning procedures, 89–90
emergency situations, 178
filtering face pieces, 86*t*
NIOSH-approved respirators, 44, 86*t*, 86*f*
requirements, 71, 85, 91
restrictions code, 83*f*
self-contained breathing apparatus (SCBA), 86*t*, 87*f*
tight-fitting respirators, 85, 86*t*
respiratory exposure
first aid procedures, 177–178
process of, 70*f*, 71
restricted-entry intervals
communication planning, 96, 98
notification requirements, 51–52
pesticide label statement, 45*f*, 47
pesticide selection, 157
safety training, 79
warning signs, 69*f*
restricted materials permit, 51
restricted-use pesticide, label sample, 43–47

restricted-use statement, pesticide label, 43, 44*f*
rinsing procedures
application equipment, 106, 122, 165
pesticide containers,103, 104*f*
rodents, 5, 12, 22–23, 32*t*
See also selection factors, pesticides
roller pumps, 112
rope wick applicators, 118*t*, 119*t*, 120*t*
routes of pesticide exposure, 30, 50, 68, 70–72
runoff hazards
cleanup cautions, 106
controlling, 166
influences, 56, 58
irrigation practices, 61
process of, 57, 68
storage site cautions, 102
rushes, 14

S

Safety Data Sheets
display requirements, 51–52
examples with explanations, 48–50, 56–57*f*
safety glasses, 84–85
See also eyewear, protective
safety procedures, summaries
application methods, 104–106
application planning, 96–100
mixing and loading, 102–104
storage facilities, 101–102
training requirements, 78–79, 190–195
See also specific topics, e.g., exposure to pesticides; mixing and loading; personal protective equipment (PPE)
sanitation procedures. *See* cleaning procedures
scab diseases, 23*f*, 25
scale pests, 19
school sites, 61–62, 106
screens, filters, 110, 113, 117
SDS. *See* Safety Data Sheets
secondary pests, 2, 63
secondary poisoning, 62–63, 169
sedges, 14
selection factors, pesticides
adjuvants, 38
chemical family, 33
effectiveness summarized, 154–157
formulations, 34–38
information resources, 154
mode of action, 33–34
natural enemies protection, 62
pest target, 32
residue control, 61
resistance problems, 161–162
selectivity, 5, 156
toxicity levels, 30–31, 156
selectivity of pesticides, 5, 156
self-contained breathing apparatus (SCBA), 86*t*, 87*f*
sensitive areas, pesticide hazards, 61–62, 105–106, 166–169
sensitization to pesticides, 73
sharpshooters, 19–20
shoes, protective, 84, 88
shrubs, 32*t*
sight gauges, 130, 133
signal words
pesticide labels, 44*f*, 46
spill material storage, 180
toxicity categories, 31, 73
warning signs, 69*f*, 98, 99, 102
skin exposure
first aid procedures, 176–177
process of, 60, 70–71
skippers, 20
slugs, 21, 32*t*
See also selection factors, pesticides
slurry formulations, 36*t*
snails, 21, 32*t*
See also selection factors, pesticides
soil characteristics, pesticide action, 56, 58–59, 157
soil drenches, furrow applications, 120–121*t*
soil injection applications, 121*t*, 137, 158
solid stream nozzles, 116*t*
solubility of pesticides, 56, 57–59, 157
soluble powders, 36–37*t*, 114, 156*t*
Sphaerotheca pannosa, 24*f*
spill accidents
emergency phone numbers, 174
exposure potential, 58, 68
kit requirements, 179–180
label information, 50
planning for, 96
response actions, 180–182
Safety Data Sheet information, 50
storage facilities, 102
transport responsibilities, 100–101
spill kits, 179–180
spores, fungi, 23–24
spotted wing drosophila, 20–21
spot treatments, 118*t*
spray check devices, 132
spray-dip applications, 120*t*
spray drift, 163*t*, 164–165
See also drift events
spray droplets. *See* droplets, spray
sprayers
cleaning, 117, 121–122
maintenance, 122–123
types, comparisons, 118–120*t*
See also application equipment; calibration methods; *and specific types, e.g.,* backpack sprayers, *and specific components, e.g.,* tanks
spray patterns, nozzle comparisons, 115–116*t*
spray swath, measuring, 134–138
spray-to-wet treatments, 118*t*
stability information, Safety Data Sheets, 50
stainless steel equipment
nozzles, 114
pumps, 112
tanks, 111
statement of use classification, 43, 44*f*
stolen pesticides, 182
storage of application equipment, 117
storage procedures, pesticides
containers, 68–69, 71, 72, 102, 103
pesticide label directions, 45*f*, 47
posting storage warning signs, 99*t*
protecting children, 68–69, 71
protective equipment, 88, 89–90
Safety Data Sheet information, 50
safety guidelines summarized, 101–102
storage locations on facility map, 175
strainers, 110, 113, 117
strip sprays, 136, 137*f*

sucking lice, 17
suits, chemical-resistant, 82
summer annuals, 12, 13*f*
supplemental labels, 42
supplied-air respirators, 86*t*, 177
surface water contamination, 57, 58–59, 61, 105–106, 165–166
swallowed pesticides, first aid, 178
swath width
  consistency importance, 158
  measuring, 134–138, 141, 145
symptoms. *See* exposure to pesticides
systemic pesticides, 34
system monitors and controllers, 112, 115*t*, 116*t*, 118*t*, 149–150

T

tailwater collection, 61
tanks
  cleaning procedures, 121–122
  filling procedures, 104
  inspection procedures, 111
  measuring capacity, 129–130
  system diagram, 110*f*
  types of, 111
Taphrina spores, 23*f*
target pests
  label requirements, 45*f*, 46–47
  pesticide selection examples, 32
temperature hazards, 105–106, 165
temperature inversions, 105*t*, 106*f*, 163*t*, 165
testing for toxicity, 30, 73
theft of pesticides, 182
thermoplastic tanks, 111
thrips, 18
ticks, 17
timing considerations
  drift prevention, 166
  nontarget species protection, 167
  pest life cycle, 158
  phytotoxicity prevention, 167
  pollinator protection, 62
  residue control, 60
  weather forecasts, 166
  See also notification requirements; restricted-entry intervals
toxicity, pesticide, 30–31, 46, 50, 73–74, 156
training requirements
  California Code of Regulations, 188
  personal safety, 69, 74, 78–79
  recordkeeping forms, 190–195
  respirator use, 85
  transport of pesticides, 101
transport guidelines, 50, 68, 100–101
traps, 5
travel speed, during application
  consistency importance, 158
  granule applicators, 142, 146
  liquid applicators, 130, 131*f*, 141
treehoppers, 19–20
trees, 32*t*
triazines, 33*t*
trigger pump sprayers, 118*t*, 119*t*
triple-rinsing procedures, 103, 104*f*
true bugs, 18–19
tungsten carbide nozzles, 114

U

ultra-low-volume applications, 36*t*, 118*t*

V

vapor drift, 163*t*
  See also drift events
Venturia inaequalis spores, 23*f*
vertebrate pests, 12, 21–23
Vicente's Vineyards, 2–3
viruses, 26–27
viscosity, droplet size, 165
volatile organic compounds (VOC), 156
volatility of pesticides, 56–57

W

WARNING signal
  pesticide labels, 31, 46
  sign posting, 69*f*, 98, 99, 102
  toxicity category, 31, 156
warning signs, 69*f*, 98, 99, 102
warranty information, 45*f*, 47
wasp predators, 3
water-dispersible granules, 35*t*, 156*t*, 159
water-resistant/waterproof clothing, 80, 84
water-soluble bags or packets, 37*t*, 92, 102–103
water-soluble concentrates, 36*t*, 159
water supplies, drinking, 61, 169
water supply, application equipment, 117, 124
weather conditions
  assessment tools, 163–164
  hazards summarized, 105*t*, 106*f*
  pest development, 154
  pesticide selection, 156–157
  recalibration of applicators, 142
  See also drift events
weeds, 12–16, 32*t*
  See also selection factors, pesticides
wells, protection, 59, 102, 103, 166
wettable powders
  dilution calculations, 147*f*, 149*f*, 150*f*
  equipment wear, 112, 114
  mixing procedures, 159
  types of, 35–36*t*, 92, 156*t*
whiteflies, 19
wildlife. *See* birds; mammal pests; nontarget species
winds, drift potential, 57, 58–59, 105*f*
winter annuals, 12, 13*f*, 14–15
wiper applications, 119*t*
work clothes, 80–84, 87–88
  See also personal protective equipment (PPE)
Worker Protection Standards, 42, 45*f*, 47
worker rights, training requirements, 79

X

Xanthomonas vesicatoria, 24*f*

Y

yellow foxtail, 12, 15
yellow nutsedge, 12, 14
yellow starthistle, 16